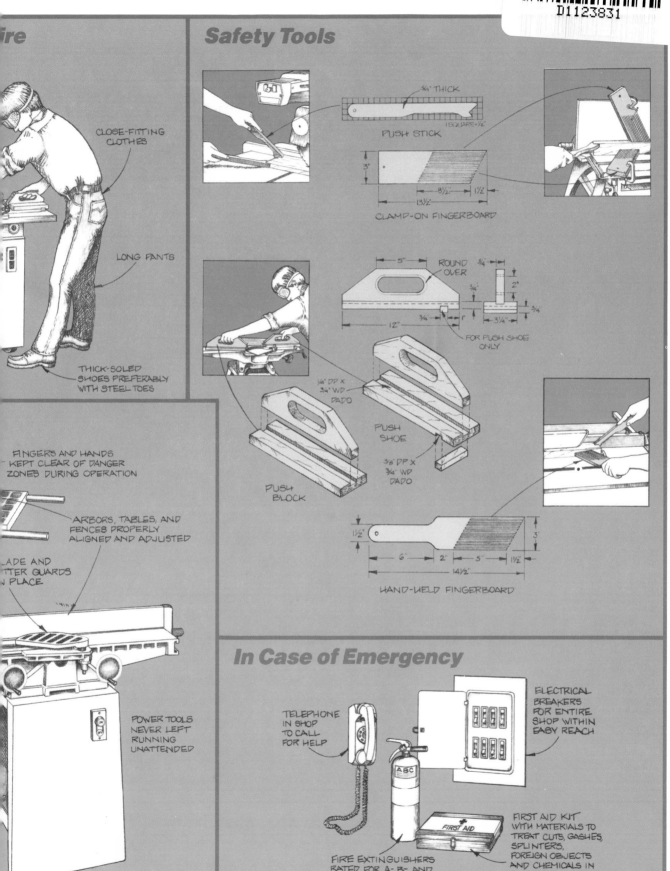

ire

CLOSE-FITTING CLOTHES

LONG PANTS

THICK-SOLED SHOES PREFERABLY WITH STEEL TOES

FINGERS AND HANDS KEPT CLEAR OF DANGER ZONES DURING OPERATION

ARBORS, TABLES, AND FENCES PROPERLY ALIGNED AND ADJUSTED

LADE AND TTER GUARDS N PLACE

POWER TOOLS NEVER LEFT RUNNING UNATTENDED

Safety Tools

¾" THICK

1 SQUARE = ½"

PUSH STICK

3"

8½" 1½"

13½"

CLAMP-ON FINGERBOARD

5" ROUND OVER ¾"

¾" 2"

12" ¾ 1" 3¼" ¾

FOR PUSH SHOE ONLY

¼" DP X ¾" WD DADO

PUSH SHOE

PUSH BLOCK

⅜" DP X ¾" WD DADO

1½" 3"

6" 2" 5" 1½"

14½"

HAND-HELD FINGERBOARD

In Case of Emergency

TELEPHONE IN SHOP TO CALL FOR HELP

ELECTRICAL BREAKERS FOR ENTIRE SHOP WITHIN EASY REACH

ABC

FIRST AID

FIRE EXTINGUISHERS RATED FOR A-, B-, AND C- CLASS FIRES

FIRST AID KIT WITH MATERIALS TO TREAT CUTS, GASHES, SPLINTERS, FOREIGN OBJECTS AND CHEMICALS IN EYES, AND BURNS.

·BUILD·IT·BETTER·YOURSELF·
WOODWORKING PROJECTS

Display Cases, Frames, and Shelves

Collected and Written
by Nick Engler

Rodale Press
Emmaus, Pennsylvania

If you have any questions or comments concerning this book, please write:
Rodale Press
Book Reader Service
33 East Minor Street
Emmaus, PA 18098

Series Editor: William H. Hylton
Managing Editor/Author: Nick Engler
Editor: Roger Yepsen
Copy Editor: Mary Green
Graphic Designer: Linda Watts
Graphic Artists: Mary Jane Favorite
            Chris Walendzak
            Christine Vogel
Photography: Karen Callahan
Cover Photography: Mitch Mandel
Cover Photograph Stylist: Janet C. Vera
Proofreader: Hue Park
Typesetting by Computer Typography, Huber Heights, Ohio
Interior Illustrations by O'Neil & Associates, Dayton, Ohio
Endpaper Illustrations by Mary Jane Favorite
Produced by Bookworks, Inc., West Milton, Ohio

Library of Congress Cataloging-in-Publication Data

Engler, Nick.
    Display cases, frames, and shelves/collected and written by Nick Engler.
        p.   cm. — (Build-it-better-yourself woodworking projects)
    ISBN 0–87857–843–9 hardcover
    ISBN 0–87857–844–7 paperback
    1. Cabinet-work. 2. Shelving. (Furniture) I. Title. II. Series:
Engler, Nick. Build-it-better-yourself woodworking projects.
TT197.E53 1989
684.1'6 — dc20   89–10876
                                                               CIP

Distributed in the book trade by St. Martin's Press

12 14 16 18 20 19 17 15 13   hardcover
    4 6 8 10 9 7 5 3   paperback

Contents

Pack Rats   1

PROJECTS

Glass-Front Wall Cabinet   2
Country Shadowboxes   10
Quilt Rack   16
One-Board Cases   23
Pigeonholes   28
Museum Table   34
Quick-and-Easy Picture Frames   43
Watch Case   52
Curio Cabinet   57
Stackable/Hangable Shelves   70
Glass-Top Display Stand   74
Corner Showcase   82
Multiple Picture Frame   88
Corner Cupboard   94
Round and Oval Frames   100
Tool Display Cabinet   107

TIPS AND TECHNIQUES

Mat-Cutting Jig for Picture Frames   51
Step-by-Step: Cutting Dovetails by Hand   121

Credits   123

Pack Rats

While many anthropologists contend that humans evolved from apes, another theory proposes that our ancestors were pack rats. There's overwhelming evidence to support this hypothesis — just open up the nearest closet and take an inventory.

Most of us have collections of one sort or another. Although we make fun of our penchant for saving things — even feel guilty about it from time to time — collections serve an important emotional and spiritual function. They connect us with our world, keep us aware and alert and interested. The simple pleasure of walking along a seashore becomes an adventure when you know what sorts of invertebrates inhabit the coastal waters. A stroll becomes a quest as you look for new shells to add to your collection or more perfect examples of other shells that you already have.

A Place for Everything...

Instead of feeling guilty, address the problem: Collections create clutter. When you save things, they begin to pile up and overwhelm you. The time-honored solution for getting rid of this clutter (without getting rid of the collection) is simple: *Organize* the things you're saving.

Simply put, create a place for them — a case, a frame, or a set of shelves. Making a project like this carries its own emotional gratification. Within a case, a collection suddenly becomes more than the sum of its parts. It takes on a new importance in your life, and saving things becomes a satisfying pastime instead of an annoying habit.

Cases and frames also reduce the work and the worry of a collection. They can protect your valuables from dust, accidents, or children. They also prevent things from being lost or misplaced. You spend more time enjoying your collection and less time maintaining it.

...And Everything in Its Place.

As you set out to build a place for your treasures, remember this: It's just as important that the case fit the collection as the collection fit the case. To be appreciated fully, different items must be stored and displayed in different ways. Consider these three attributes:

Size — Obviously, the size of your collection will affect the size of the case or frame. But what about the size of the items themselves? Are they all the same? Or is there a range of sizes? Display units like the *Quilt Rack* and *Country Shadowboxes* are made for things that are pretty much the same proportions. However, if the dimensions differ greatly from item to item, build something more versatile, like the *Tool Display Cabinet* or *Multiple Picture Frame*.

View — How should the items in your collection be viewed? More important, how much of each item do you need to see? When displaying a photograph, you need to see only one side — an ordinary picture frame serves nicely. The same is true for watches, keys, and many other small items. The *Watch Case* or the *Glass-Front Wall Cabinet* will adequately display these. But some objects, such as carved birds or dolls, must be seen from all sides to be appreciated. These should be displayed in or on units like the *Curio Cabinet* or *Glass-Top Display Stand*. And other items fall somewhere in the middle — you want to view them from several different angles, but you don't need to see every surface. You might display these in the *Museum Table* or *Corner Showcase*.

Access — Do you use the items in your collection often? Do they have to be taken out of their case and handled to be fully enjoyed? You don't need to — or want to — handle photographs. You can view them mounted in a frame, behind glass and out of reach. But if you have a collection of old tools, you want to hold these — maybe even use them from time to time — to gain an intimate understanding of them. The *Tool Display Cabinet, Quilt Rack*, and *Corner Cupboard* all let you reach objects easily.

This book contains plans for cases, frames, racks, and shelves that will hold all sizes and kinds of keepsakes and treasures in various ways. Page through it for ideas before you start to build, then mix and match the options to create a place that fits your collection perfectly.

Glass-Front Wall Cabinet

Hanging cabinets offer an advantage over most other shelving and display units: They make use of otherwise wasted space. Hang them over other furniture, or set them on tables and counters. Place them wherever you have a vacant wall, even if the floor area underneath is already occupied.

There are other advantages, too. By positioning them at a comfortable height, you can reach the shelves easily. The glass doors let you see and enjoy the contents.

Yet the items inside remain enclosed, protected from dust and too much handling.

This glass-front cabinet was designed and built by W. R. (Rick) Goehring, a professional cabinet-maker in Gambier, Ohio. Rick specializes in designing and building "country derived" pieces such as this. He studies the country forms, particularly those of ethnic groups such as the Shakers, Moravians, and Amish, then builds his own interpretations.

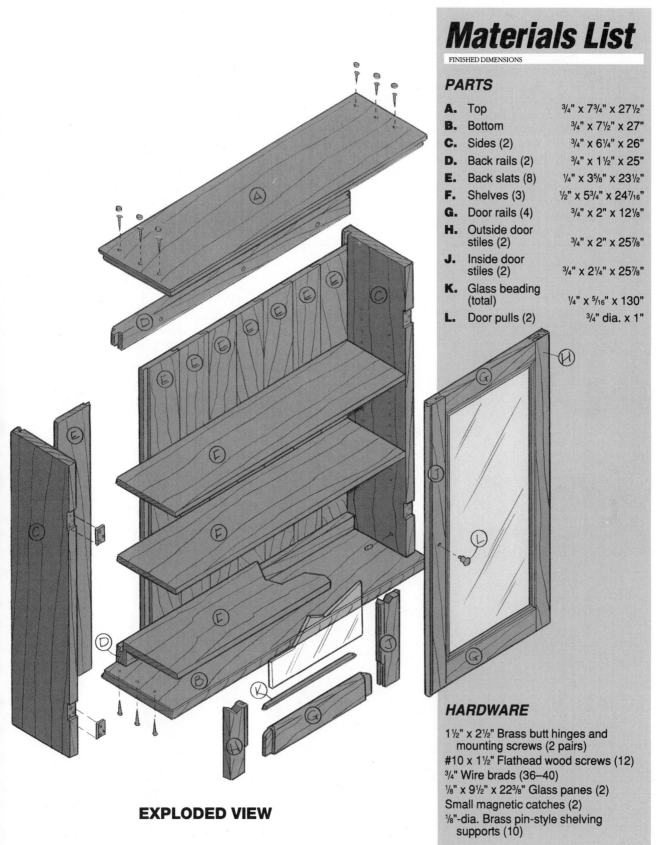

EXPLODED VIEW

Materials List

FINISHED DIMENSIONS

PARTS

A.	Top	3/4" x 7 3/4" x 27 1/2"
B.	Bottom	3/4" x 7 1/2" x 27"
C.	Sides (2)	3/4" x 6 1/4" x 26"
D.	Back rails (2)	3/4" x 1 1/2" x 25"
E.	Back slats (8)	1/4" x 3 5/8" x 23 1/2"
F.	Shelves (3)	1/2" x 5 3/4" x 24 7/16"
G.	Door rails (4)	3/4" x 2" x 12 1/8"
H.	Outside door stiles (2)	3/4" x 2" x 25 7/8"
J.	Inside door stiles (2)	3/4" x 2 1/4" x 25 7/8"
K.	Glass beading (total)	1/4" x 5/16" x 130"
L.	Door pulls (2)	3/4" dia. x 1"

HARDWARE

1 1/2" x 2 1/2" Brass butt hinges and mounting screws (2 pairs)
#10 x 1 1/2" Flathead wood screws (12)
3/4" Wire brads (36–40)
1/8" x 9 1/2" x 22 3/8" Glass panes (2)
Small magnetic catches (2)
1/8"-dia. Brass pin-style shelving supports (10)

1 **Select and cut the stock.** To make this project, you need approximately 18 board feet of 4/4 (four-quarters) cabinet-grade lumber. You may use any wood you like, even mixing different hardwoods. Rick sometimes uses contrasting woods as part of his design. Walnut and cherry look good together, as do walnut and maple, oak and ash, cherry and birch, and rosewood and teak. Whatever woods you use, choose the clearest, *straightest* stock you can find for the door frame parts. If the rails or stiles are distorted at all, the doors may not close properly.

Plane 10 board feet of lumber to ¾" thick and 4 board feet to ½". Resaw the remaining 4 board feet in half on a band saw, then plane it to ¼" thick. Cut all the parts to size *except* the glass beading. (Set the ¼"-thick stock for the beading aside until later.) If you have some ¾"-thick stock left over, make several extra rails and stiles to use in checking the tool setups.

TRY THIS! To get the doors to fit perfectly, you may want to cut the rails and stiles 1/16" longer than specified in the Materials List. Build the doors oversize, then plane and sand them to fit the case.

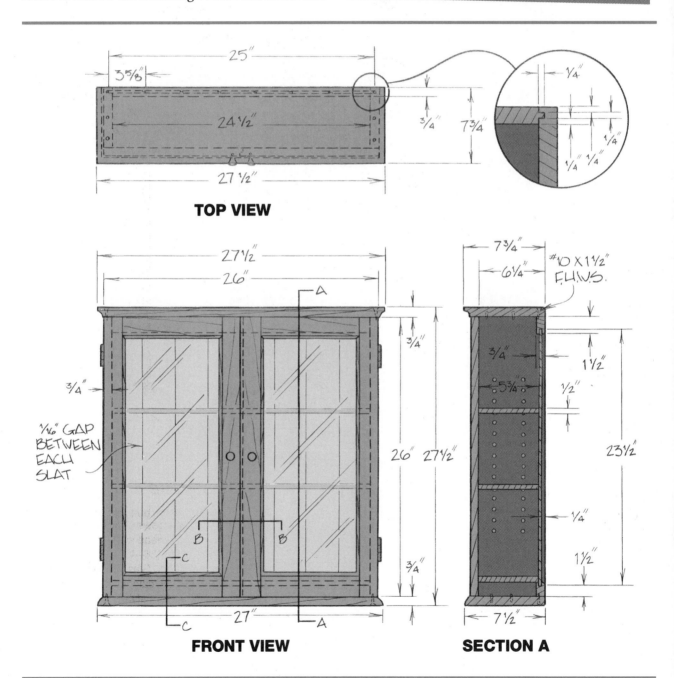

TOP VIEW

FRONT VIEW

SECTION A

2 Shape the top, bottom, and door frame.

Several of the parts in the cabinet must be shaped. You can cut these shapes with either a shaper or a table-mounted router:

- Cut a cove in the front edge and ends of the *top*, as shown in the *Top Edge Profile*.
- Round over the front edge and ends of the *bottom*, as shown in the *Bottom Edge Profile*.
- Cut a small bead in the front face of the *right inside stile*, as shown on *Section B*.
- Cut a ⁵⁄₁₆"-wide ogee in the inside edges of all the *rails* and *stiles*, as shown in *Sections B* and *C*.

Sand the shaped edges to remove any mill marks or burn marks.

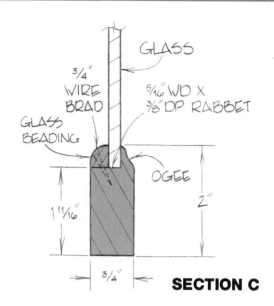

SECTION C

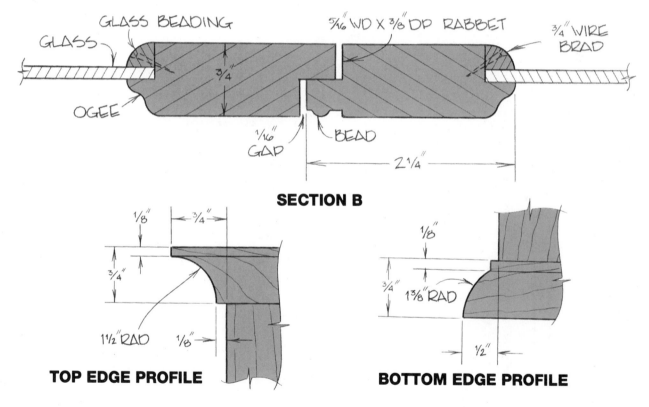

SECTION B

TOP EDGE PROFILE

BOTTOM EDGE PROFILE

3 Shape and cut the glass beading.

While you're set up to do shaping, make the glass beading. Round over the edge of a piece of ¼"-thick stock, as shown in the *Glass Beading Profile*. Then rip the ⁵⁄₁₆"-wide beading from the edge. Repeat, until you have made all the beading you need. If you have enough stock, make extra in case some beading splits when you nail it in place.

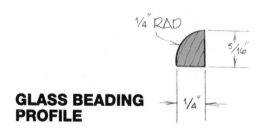

GLASS BEADING PROFILE

4 Cut the joinery in the sides, back rails, and back slats.

Drill ⅛"-diameter, ⅜"-deep holes in the inside faces of the sides, as shown in the *Side Layout*. (These holes will hold the shelving supports for the adjustable shelves.) Mortise the front edges for the butt hinges. Using a dado cutter or router, cut ¼"-wide, ¼"-deep grooves in the sides, near the back edges to hold the back slats.

Cut or rout ¼"-wide, ¼"-deep grooves in the inside edges of the back rails, as shown in the *Back Rail and Slat Detail*. Also make ¼"-wide, ¼"-long tongues in each end of the rails, as shown in the Top View. These tongues must fit in the grooves in the sides. (See Figure 1.)

Cut or rout ½"-wide, ⅛"-deep rabbets in *both* edges of *six* back slats and *one* edge of the remaining *two*. These two slats will become the left- and right-most slats.

1/When cutting tenons with a dado accessory or table-mounted router, use a miter gauge to guide the back rail. Also use a stop block to help position the rail in the miter gauge. By using a stop block instead of the fence, you prevent the wood from binding and kicking back.

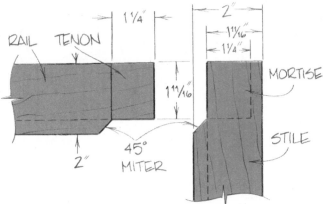

MORTISE-AND-TENON LAYOUT

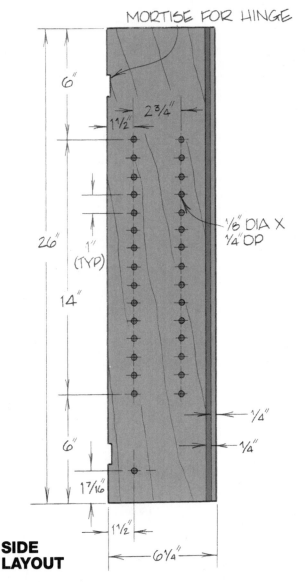

SIDE LAYOUT

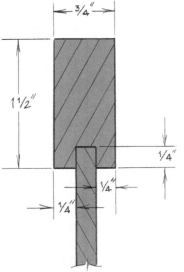

BACK RAIL AND SLAT DETAIL

5 **Assemble the case.** Finish sand the parts you've made so far, and dry assemble them to test the fit. When you're satisfied they fit properly, place the slats in the grooves in the back rails. *Don't* glue them in place, just let them float. Then glue the back rail tenons in the grooves in the sides. Once again, *don't* glue the slats.

TRY THIS! In the assembled cabinet, the back slats will expand and contract with changes in humidity. When they contract in the dry part of the year, the gap between the back slats will widen, revealing a little of the wood in the lapping rabbets. If you leave this wood natural, these gaps will be very noticeable. To prevent this, apply a dark stain to the rabbets before you assemble the cabinet. Be careful not to stain any other parts of the wood. The gaps will still be there part of the year, but dark gaps are much less conspicuous than light-colored ones.

Attach the top and bottom to the assembly with glue and flathead wood screws. Counterbore and countersink the screws, so you can cover the heads with wood plugs. Cut these plugs out of scraps of the same hardwood used to build the cabinet, matching the color and the grain as closely as possible. Glue the plugs in the counterbores, let the glue dry, then sand the plugs flush with the wood surface.

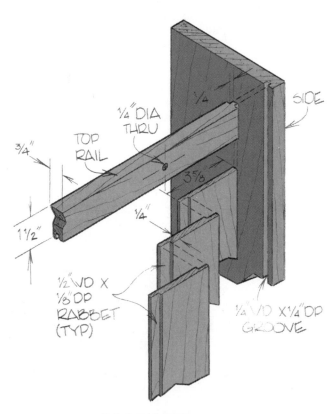

BACK ASSEMBLY DETAIL

---

6 **Cut the door joinery.** Using a table-mounted router and a ¼"-straight bit, rout a ¼"-wide, 1⁹⁄₁₆"-deep, 1¹¹⁄₁₆"-long mortise in each end of the stiles. (See Figure 2.) Square the blind end of this mortise with a chisel. Then cut a ⁵⁄₁₆"-wide, ⅜"-deep rabbet to hold the glass in the inside edge of all the rails and stiles, opposite the ogee shape. (See Figure 3.) Also cut a ⁵⁄₁₆"-wide, ⅜"-deep rabbet on the *inside* face of the *right inside stile,* opposite the bead, and on the *outside* face of the *left inside stile,* as shown in *Section B.*

2/When routing a blind mortise, use a stop block to halt the cut. Feed the stock so the rotation of the bit helps to hold it against the fence. Make the mortise in several passes, routing just ⅛" to ¼" deeper with each cut.

3/Cut a rabbet in the back face of all the rails and stiles, opposite the ogee. Once again, make this joint in several passes.

Cut a ¼"-wide, 1¼"-long tenon in the full width of each rail. (See Figure 4.) Using a band saw or dovetail saw, trim 1¹¹⁄₁₆" of the ogee shape from each end of the stiles and 1¼" from each end of the rails. (See Figure 5.) Then miter the ends of the ogee shape at 45°, as shown in the *Door Joinery Detail*. Dry assemble the doors to check the fit of the mortise-and-tenon joints. The mitered ends of the ogees should come together like a picture frame.

4/Make the tenons on the ends of the door rails in a similar manner as those on the back rails. However, since these tenons are much longer, make them in several passes.

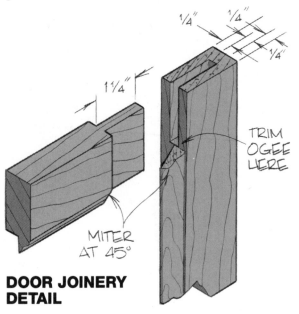

DOOR JOINERY DETAIL

TRIM OGEE HERE

MITER AT 45°

5/For the door frame parts to fit together properly, you must trim and miter the ogees. To get a good fit on both the miters and the mortise-and-tenon joints, cut the ogees a little wide of your marks. Then trim the wood with a chisel until the pieces fit perfectly.

---

7 Assemble and hang the doors. Lightly sand the rails and stiles, then glue them together. After the glue dries, sand the joints clean and flush. Fit the doors to the case, planing them until there's just a ¹⁄₁₆" gap between each door and a ¹⁄₁₆" gap at the top and bottom. Be careful to plane the outside edges of the door *straight*.

Place the doors in the cabinet and mark the position of the hinges. Using a chisel, mortise the back face of the door frames for the hinges. Screw the hinges to the doors, then to the cabinet. Test the action of the doors, and adjust the position of the hinges if necessary. Once the doors swing open and shut freely, without binding, install magnetic catches inside the cabinet.

---

8 Finish the completed cabinet. Test fit the shelves in the cabinet. Then remove the shelves and the doors from the case. Detach all the hardware and do any necessary touch-up sanding. Apply a finish to all the wooden parts and assemblies, including the glass beading. Be sure to apply as many coats of finish inside the case as outside — this will keep it from distorting.

When the finish has dried, place the glass in the door frames. Cut the glass beading to fit inside each door frame, mitering the corners. Nail the beading in place with wire brads. Set the heads of the brads and, if necessary, touch up the finish on the beading.

Hang the doors on their hinges again. Also reinstall the catches, shelving supports, and shelves.

9   *Hang the cabinet on a wall.*   To hang the cabinet on a wall, locate the studs in the section of wall that the cabinet will span. Drill three or four ¼"-diameter holes through the top back rail, evenly spaced along the board. Wherever you can, align these holes with the studs.

Hold the cabinet on the wall and mark the wall through the holes. Drill ³⁄₁₆"-diameter pilot holes in the wall at those marks that are over a stud. Drill ⁵⁄₁₆"-diameter holes and install Molly anchors at those marks that aren't. Have a helper hold the cabinet in place while you drive #14 roundhead wood screws into the studs and ¼" bolts into the anchors.

Note: If you hang this cabinet on a masonry wall, secure it with three or four expandable lead anchors and ¼" lag screws.

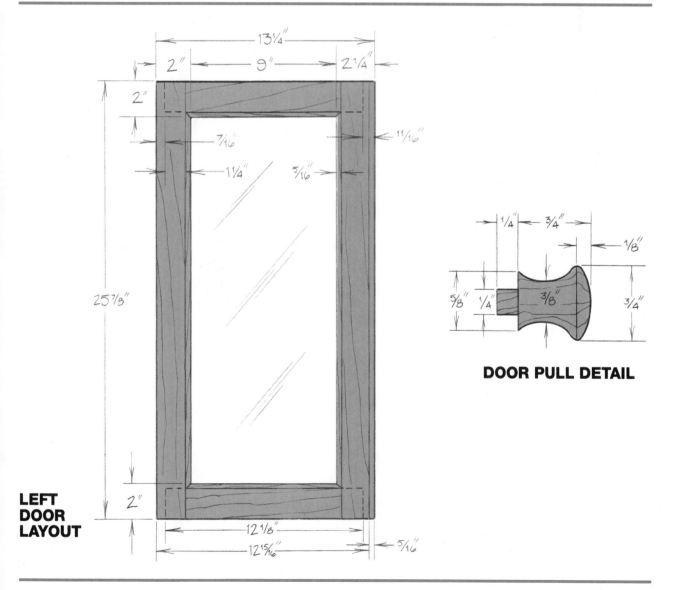

LEFT DOOR LAYOUT

DOOR PULL DETAIL

Country Shadowboxes

Shadowboxes are picture frames for *three-dimensional* objects. Place an object in a shadowbox, and it becomes the centerpiece of a larger design. The box focuses your attention on the article, enhancing its beauty.

These particular shadowboxes are built especially for country collectibles.

Depending on what you want to display, you can build them with one, two, three, or more compartments. You can also choose the shape of the opening that best frames your trinkets. Shown are three possibilities: a heart, an arch, and a traditional quilting pattern called an "Ohio Star."

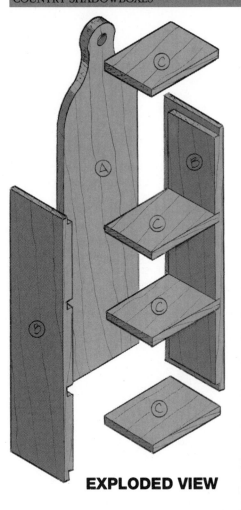

EXPLODED VIEW

Materials List

FINISHED DIMENSIONS

PARTS

Three-Compartment Shadowbox

A.	Back	½" x 5½" x 22"
B.	Sides (2)	½" x 4" x 17"
C.	Shelves (4)	½" x 3½" x 5½"
D.	Front	¼" x 6" x 17"

Two-Compartment Shadowbox

A.	Back	½" x 5½" x 15½"
B.	Sides (2)	½" x 4" x 11½"
C.	Shelves (3)	½" x 3½" x 5½"
D.	Front	¼" x 6" x 11½"

EXPLODED VIEW

One-Compartment Shadowbox

A.	Back	½" x 5½" x 7"
B.	Sides (2)	½" x 4" x 6"
C.	Shelves (2)	½" x 3½" x 5½"
D.	Front	¼" x 6" x 6"

HARDWARE

1" Wire brads (18–40 per shadowbox)

EXPLODED VIEW

1

Select and cut the stock. Choose a stable wood to make these boxes, one that doesn't expand and contract too much with changes in the weather. The reason stable materials are necessary is that no matter how you assemble these shadowboxes, the wood grain will be opposed at some joints. The grain direction of the back and the front is parallel, so these two parts will expand and contract in unison. But if you run the grain of the shelves front-to-back, it will be perpendicular to the grain of the sides; if you run it side-to-side, it will be perpendicular to the grain of the front. Either way, one part will prevent another from moving.

By the way, it's best to run the shelf grain front-to-back, so it opposes the grain of the sides. The sides are ½" thick and better able to withstand this opposition than the ¼"-thick front. If the shelf grain opposes the front grain, the front may split.

The boxes shown are made from poplar, an extremely stable hardwood. Cherry, maple, and birch are also good choices. Or, if you have some scraps of cabinet-grade ½" plywood, use this to make the shelves *only*. (Make the other parts from solid wood.) Plywood shelves remove the expansion/contraction problem altogether, without detracting from the appearance of the boxes.

To make a three-compartment shadowbox, you need approximately 3½ board feet of 4/4 (four-quarters) lumber. (One- and two-compartment boxes require proportionately less.) Resaw 1 board foot of material to make ¼"-thick stock, and plane the rest to ½" thick. Then cut the parts to the sizes in the Materials List.

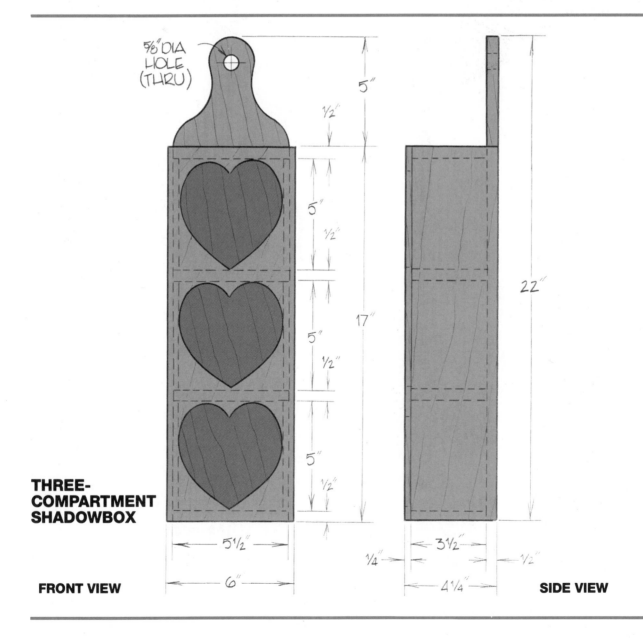

THREE-COMPARTMENT SHADOWBOX

FRONT VIEW

SIDE VIEW

2 Cut the joinery.

Using a table-mounted router or a dado cutter, cut ½"-wide, ¼"-deep dadoes and rabbets in the sides, as shown in the *Front* Views and *Side Views*. Also cut a ½"-wide, ¼"-deep rabbet along the back edge of each side.

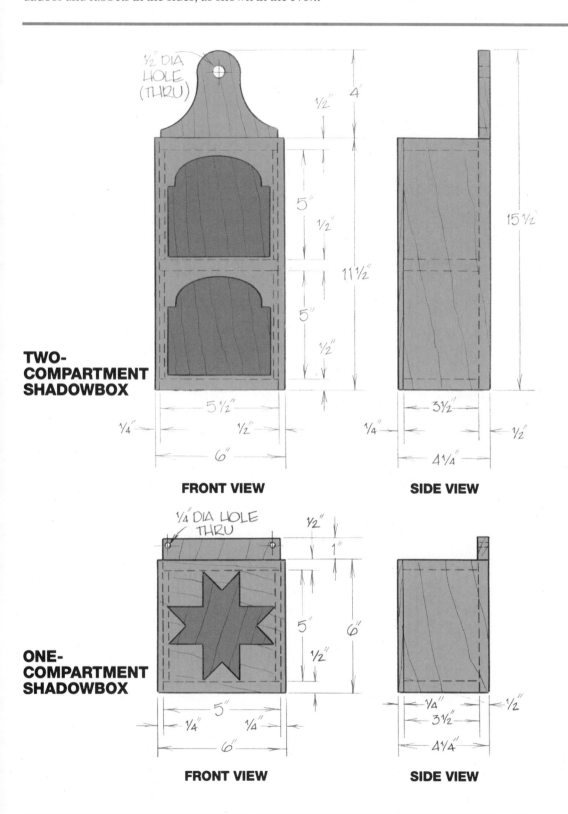

TWO-COMPARTMENT SHADOWBOX

FRONT VIEW SIDE VIEW

ONE-COMPARTMENT SHADOWBOX

FRONT VIEW SIDE VIEW

3

Shape and drill the back. Decide which back pattern you want to use for the shadowbox. You can use any pattern with any size box, or you can leave the back plain, as shown on the one-compartment shadowbox.

If you choose to cut a pattern, enlarge it and trace it on the stock. Cut it with a band saw or saber saw. Sand the sawed edges, then drill a mounting hole near the top edge. If you leave the back plain, drill two mounting holes, one near each top corner.

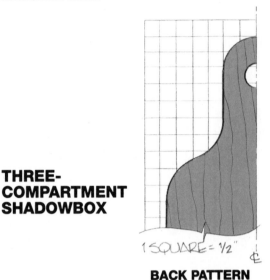

**THREE-
COMPARTMENT
SHADOWBOX**

1 SQUARE = ½"

BACK PATTERN

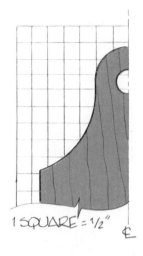

**TWO-
COMPARTMENT
SHADOWBOX**

1 SQUARE = ½"

BACK PATTERN

4

Make the cutouts in the front. Mark the front stock to show where the shelves and sides will join it. Choose which cutout pattern you want to use. As with the back patterns, you can use any cutout with any size box. Enlarge the pattern and trace it on the front. Carefully center the cutouts in the open areas between the marks for the shelves and sides. Remem-

ber, *you should not be able to see the front edges of the sides or shelves* after you assemble the shadowbox.

Drill a hole approximately ¾" in diameter through each cutout pattern, inside the lines. Insert the blade of a saber saw, jigsaw, or scroll saw through this hole. Saw to the pattern line, then cut out the waste. Sand the sawed edges of the cutout.

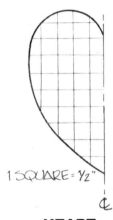

1 SQUARE = ½"

**HEART
CUTOUT
PATTERN**

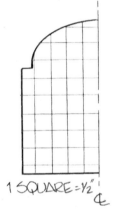

1 SQUARE = ½"

**ARCH
CUTOUT
PATTERN**

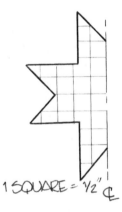

1 SQUARE = ½"

**OHIO STAR
CUTOUT
PATTERN**

5 Assemble the shadowbox.

Assemble the shadowbox. Dry assemble the shadowbox to make sure the parts fit together properly. When you're satisfied they do, finish sand all parts, being careful not to round adjoining edges. Assemble the box with glue and wire brads.

Let the glue dry, then sand all joints clean and flush. Set the brads and cover the heads with wood putty or stick shellac. Sand the putty or shellac flush with the wood surface.

TRY THIS! If you don't want to see the heads of the brads, use a blind nail plane to hide them. Wherever you need to drive a brad, cut a tiny sliver or curl of wood with the plane. Hammer the brad through the hollow left by the curl, then glue the curl back in place. After the glue dries, sand the area clean. The brad will lie hidden just under the surface of the wood.

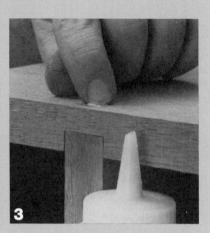

6 Finish the shadowbox.

Finish the shadowbox. Do any necessary touch-up sanding, then apply a finish to the completed box. The boxes shown are painted a bright color on the inside and finished with clear tung oil on the out-side. If you finish your shadowboxes in a similar manner, choose a color that will complement the objects in the box. You don't want colors to clash, nor do you want the objects to blend into the background.

TRY THIS! When painting the insides of the shadowboxes, cover the inside edges of the cutouts with masking tape. This will keep them clean.

Quilt Rack

*B*ack when cloth goods were woven from strictly organic fibers, storage was hard on quilts and other bedding. Closed up in a chest or a drawer during warm, humid weather, the fabric might mildew or be eaten by moths. It could also acquire an unpleasant, musty odor. To prevent this, folks took out their stored bedding every so often and hung it on a rack to air. They also rotated the bedding, exchanging the quilts on the rack for those in the drawers every month or so.

Synthetic fibers and air conditioning have eliminated the need for this practice. Quilt racks, however, continue to be useful. Many collectors use them to preserve and display old country quilts, afghans, and lap robes. The racks are also an excellent place to hang towels in country bathrooms and kitchens.

The quilt rack shown is designed after the fashion of many early seventeenth-century pieces. The end posts and the racks are turned, while the legs, arms, and stretcher are shaped with a band saw or scroll saw. All the parts are assembled with dowels or round mortises and tenons. ✱

Materials List

PARTS

A. End posts (2)   1¾" x 1¾" x 28"
B. Middle rail   ¾" dia. x 24½"
C. Outer rails (2)   ¾" dia. x 25½"
D. Arms (4)   ¾" x 4" x 8½"
E. Legs (4)   ¾" x 4¾" x 12"
F. Stretcher   ¾" x 3½" x 23½"
G. Dowels (20)   ⅜" dia. x 1½"

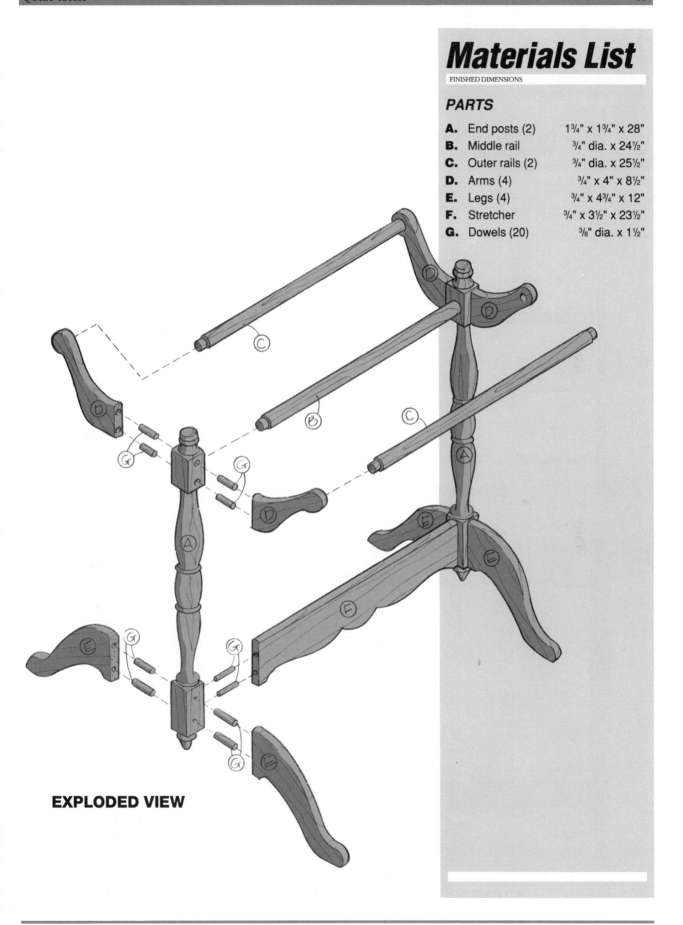

EXPLODED VIEW

1

Select and prepare the stock. A country cabinetmaker might have built a quilt rack from almost any domestic hardwood, as long as it provided sufficient strength. If you're building a traditional rack, the most common choices are walnut, cherry, and maple. Also consider fruitwoods, hickory, and birch. Avoid softwoods such as cedar and pine. Long, slender spindles turned from softwoods may begin to sag or bow after a short time.

After choosing the wood, plane it to the required thicknesses — 1¾" for the end posts, and ¾" for the arms, legs, and stretcher. Don't bother planing stock for the rails; you can make these parts from rough lumber. Cut the legs, arms, and stretcher to the dimensions shown in the Materials List. Make the end posts about 2" longer than the given dimension, so you have a little extra stock with which to mount them on the lathe. Rip 1"-wide strips from 4/4 (four quarters or 1"-thick) rough-cut lumber to make turning stock for the rails. Cut the rail stock slightly long also. You'll cut them to their proper length after you turn them.

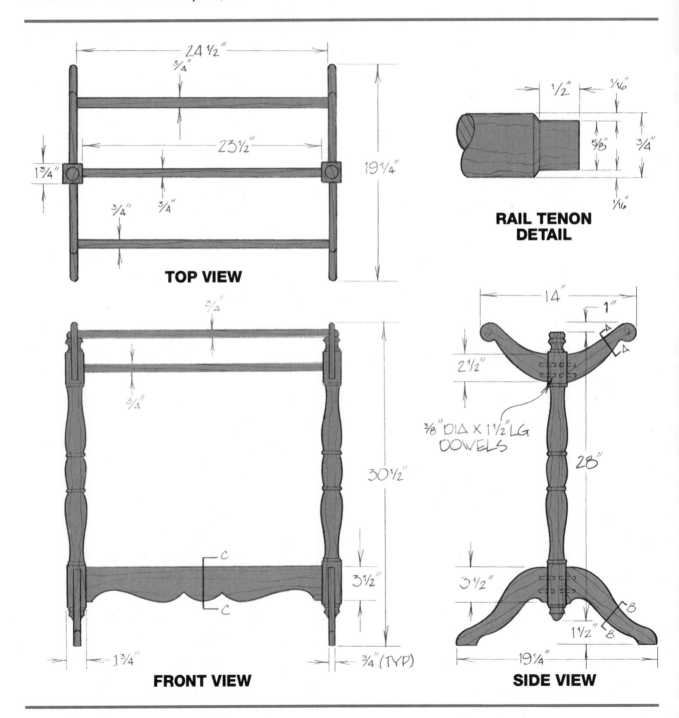

TOP VIEW

RAIL TENON DETAIL

FRONT VIEW

SIDE VIEW

2

Turn the rails. Mount a piece of rail stock on the lathe. Advance the centers until the wood is held firmly between them, but not so tightly that it begins to bow.

Starting at one end of the stock, round an 8"–10" section and turn it to $^{13}/_{16}$" in diameter. Mount a steadyrest on the lathe so it supports the section you have just turned; this will keep the spindle from whipping and bowing. Turn another 4"–5" beyond that section, and move the steadyrest 4"–5". Repeat this procedure until you have rough-turned the entire length of the stock. Then position the steadyrest in the middle of the turning and carefully work the diameter down to a uniform $^{3}/_{4}$". (See Figure 1.)

Carefully measure and mark the positions of the $^{5}/_{8}$"-diameter, $^{1}/_{2}$"-long tenons on both ends of the rail. Then turn them, as shown in the *Rail Tenon Detail*. Finish sand the turning on the lathe, but *don't* sand

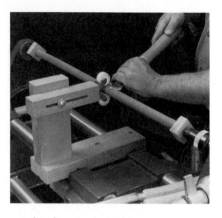

1/When turning the rails, use a steadyrest to keep the long, slender stock from whipping on the lathe. Sharp chisels and gentle pressure will also help to control whipping.

the tenons. Sanding might change their diameter, and tenons that are a little rough have a better gluing surface.

Repeat this procedure for the other two rails. Cut the waste from both ends of each turning so all three rails are the proper length.

3

Drill the round mortises and dowel holes in the end posts. Carefully measure and mark the locations of the *square* portions of one end post. Also mark the locations of the ten dowel holes and the round, $^{5}/_{8}$"-diameter mortise, as shown in the *End Post Joinery Detail*. Once you have marked one piece, lay it alongside the other. Use the first piece to locate the joinery on the second. This ensures that both pieces will be precisely the same. Drill the holes and mortises in both pieces.

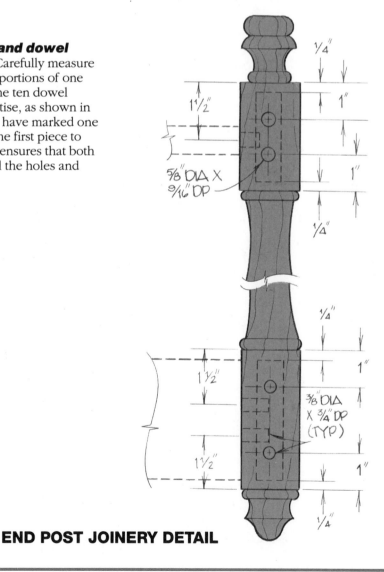

$1/4$"

1"

1"

$1^1/2$"

1"

$^{5}/_{8}$" DIA X $^{9}/_{16}$" DP

$1/4$"

$1/4$"

1"

$1^1/2$"

$^{3}/_{8}$" DIA X $^{3}/_{4}$" DP (TYP)

$1^1/2$"

1"

$1/4$"

END POST JOINERY DETAIL

4 Turn the end posts.

Turn the end posts. Not only must the joinery in both end posts be the same, but you also must turn them to the same shape. The easiest way to duplicate long turned pieces is to make a storystick. This is a simple, shopmade measuring stick marked with the dimensions of the part that you'll need to copy.

Cut a storystick from a piece of ¼"-thick scrap stock. The storystick for this turning should measure about 2" wide and 30" long. Using one long edge as the center-line, draw *half* of the end post's profile on the stick, as if you had split in down the middle. Follow the shapes shown in the *End Post Layout*. Don't worry about getting the curves of the coves and the beads just right; the shape isn't critical so long as you make both posts the same. Transfer all the dimensions on the *End Post Layout* to the drawing on the storystick.

Mount a piece of end post stock on the lathe and round the portions you want to turn. Be careful *not* to round the square sections where you've drilled the joinery! Use the point of the skew chisel to score the corners of these square areas; cleanly slicing the grain where a square section stops and a round section begins. This will keep the corners from chipping or breaking.

Lay the storystick against the post and mark the location of all the beads (convex curves) and coves (concave curves) with a pencil. Some turners use different types of pencil lines to distinguish beads from coves. They usually show the high spots (or the crests of beads) as light lines, and the low spots (the beginnings and ends of beads) as heavy lines. Shaded areas between lines are coves. A line within a shaded area indicates its low spot. (See Figure 2.)

Cut grooves to establish the diameters at the crests of the beads, low spots of the coves, and any other spots that might be helpful. (These diameters are called out on the left side of the *End Post Layout*.) To turn a groove to a specific diameter, use a parting tool and calipers. Set the calipers to the proper measurement and hold them in one hand. With the other hand, place the parting tool against the turning and slowly feed it into the wood. Monitor your progress with the calipers. As soon as they slip over the wood, stop turning. Cut grooves for each diameter along the length of the workpiece. (See Figure 3.)

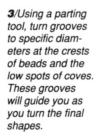

END POST LAYOUT

Measurements shown on layout: 1½" DIA, 1⅝" DIA, 1⅝6" DIA, 1⁹⁄₁₆" DIA, 1¾" SQUARE, 1¾" DIA, 1³⁄₁₆" DIA, 1¾" DIA, 1¾" DIA, 1¾" DIA, 1¾" DIA, 1¾" DIA, 1³⁄₁₆" DIA, 1¾" DIA, 1¾" SQUARE, 1⅝" DIA, 1³⁄₁₆" DIA, 1½" DIA. Other dimensions: ¾", ⅞", ¼", ⅜", 3", ¼", 6¼", ¼", 3", 28", ¼", 6¼", ¼", 1", 1", 4", ¼".

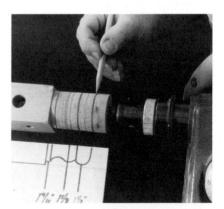

2/Mark the locations of the beads and coves on the rounded portions of the stock. A heavy line indicates the beginning or end of a bead; a light line shows the crest of the bead. Shading between lines indicates a cove.

3/Using a parting tool, turn grooves to specific diameters at the crests of beads and the low spots of coves. These grooves will guide you as you turn the final shapes.

Turn the shapes. You'll probably find it easier to make the beads first, then the coves. Use the grooves as guides: When you turn a shape down to the bottom of a groove, stop there. You know you've reached the proper diameter. Some turners draw pencil lines on the bottom of each groove as a visual aid. When they start removing a pencil line, they know to stop cutting.

As the end post takes shape, lay the storystick beside it now and then to check your progress. (See Figure 4.) Try to get the contours on the turning to match those on the stick. Don't worry about making them precisely the same; they just need to be close. As mentioned before, the shapes aren't critical.

Repeat this procedure, making a second end post. While the second post is still on the lathe, compare its shape with the first. If necessary, do a little additional turning on one post or the other (whichever appears thicker) until the two match as closely as possible.

Finish sand the turnings on the lathe. Cut off the waste on the ends of the posts and finish sand the ends.

4/As the turning takes shape, **turn off the lathe** every so often to compare the contours of the post and the storystick. Try to make them match.

TRY THIS! To make a turning as smooth as possible, sand it on the lathe *twice*. Between sandings, wipe it with a wet cloth and let it dry. The water raises the grain slightly. Sand it a second time with very fine sandpaper or steel wool to remove any loose wood fibers.

5 Cut the shapes of the arms, legs, and stretcher.

Enlarge the *Arm, Leg, and Stretcher Patterns*. Trace them onto the stock, and drill the round, ⅝"-diameter, 9⁄16"-deep mortises in the arms. Remember that each pair of arms is a mirror image of the other — drill the mortises in two arms from one side, and in the other two from the other side. Cut the shapes of the arms, legs and stretcher with a band saw or scroll saw. Sand the edges to remove any saw marks.

Round over the edges with a router and a piloted ¼" quarter-round bit. Rout both the top and bottom edges of the arms, and the top edges only of the legs and stretcher, as shown in *Section A, Section B,* and *Section C.*

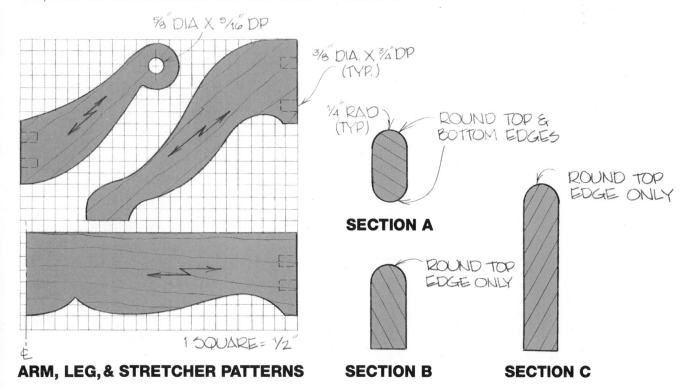

⅝" DIA X 9⁄16" DP

⅜" DIA. X ¾" DP (TYP.)

¼" RAD (TYP.)

ROUND TOP & BOTTOM EDGES

SECTION A

ROUND TOP EDGE ONLY

SECTION C

ROUND TOP EDGE ONLY

SECTION B

1 SQUARE = ½"

ARM, LEG, & STRETCHER PATTERNS

6

Drill dowel holes in the ends of the arms, legs, and stretcher. The arms, legs, and stretcher are all joined to the end posts with dowels. Since you've already drilled the dowel holes in the posts, you must make matching holes in the adjoining parts. To do this, use dowel centers to locate the holes. These are small, pointed metal inserts, available from most woodworking suppliers and some hardware stores.

To use the centers, first measure where an arm, leg, or stretcher will join the post. Then place dowel centers in the holes in the post, and press the part firmly against the post. *You must position the part precisely where it* is to join the post. The dowel centers will leave indentations in the end of the part, showing where to drill matching dowel holes. (See Figure 5.) Do this for each arm and leg and the stretcher. As you work, pay careful attention to the orientation of the round mortises. The mortises in each post must face the same direction as the mortises in its adjoining arms. Carefully mark each part so you know which post it joins and where.

Once you've marked the holes, drill them. Clamp the parts in a vise to hold them steady while you work. Use a doweling jig to help keep the drill bit straight in the stock. Fasten a drill stop to the bit to halt it when the holes are 1" deep. (See Figure 6.)

5/Dowel centers leave indentations in the end of an arm, leg, or stretcher when it's pressed against the post. To make the indentations more visible, rub the points of the centers with a lead pencil before you press the parts together.

6/A doweling jig will keep the bit square to the end of the stock as you drill the dowel holes. A drill stop fastens around the bit and keeps it from boring beyond a certain depth. If you don't have a drill stop, wrap a piece of masking tape around the bit to indicate when you should stop drilling.

7

Assemble and finish the quilt rack. Dry assemble the quilt rack to ensure that all the parts fit properly. Make any adjustments necessary to get a good fit, then disassemble the project. Finish sand the arms, legs, and stretcher.

Attach the arms and legs to the end posts with dowels and glue. Clamping these parts together presents an interesting problem: The arms and legs are odd shapes *and* they join the posts at angles. This makes them almost impossible to clamp in the usual fashion. You can manage, however, by using your hand screws creatively. Secure each arm and leg between the jaws of a hand screw. The jaws must be parallel to and about ½" away from the end where the part joins the post. Glue the part to the post, then clamp it in place by attaching another set of hand screws to the first set. (See Figure 7.) This technique is clamp intensive. If you don't own a lot of hand screws, you may have to glue up just one pair of arms or legs at a time.

After you have joined all the arms and legs to the posts, join the post assemblies with the rails and

7/To clamp the arms and legs to the posts while the glue dries, use hand screws as shown. You'll need at least four hand screws to clamp a single pair of arms or legs.

stretcher. (You can hold these parts together with bar clamps or band clamps, used in an ordinary manner.) Let the glue dry. Do any necessary touch-up sanding, then apply a finish.

One-Board Cases

Most display units are made from several pieces of wood. They have a top, bottom, sides, rails, stiles, and so on. Even a simple rectangular picture frame has four parts. The small cases shown here, however, are cut from a single board. They have been carefully sliced apart with a band saw and scroll saw, then glued together so the board appears whole.

These cases can be made from any piece of wood and adapted to hold most small items. The dark-colored knife case is cut from a chunk of walnut burl. The striped case is laminated from scraps of maple, walnut, and cherry, then hollowed-out to hold a set of darts. In addition to knives and darts, you can create one-board cases to carry or display coins, jewelry, sewing notions — almost anything, as long as it's smaller than the board you use to make the case.

Note: This project is built to fit the items you want to carry or display — there are no set sizes. You can also use a variety of hinges and catches. For this reason, there is no Materials List. ●

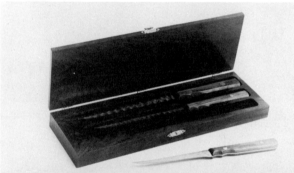

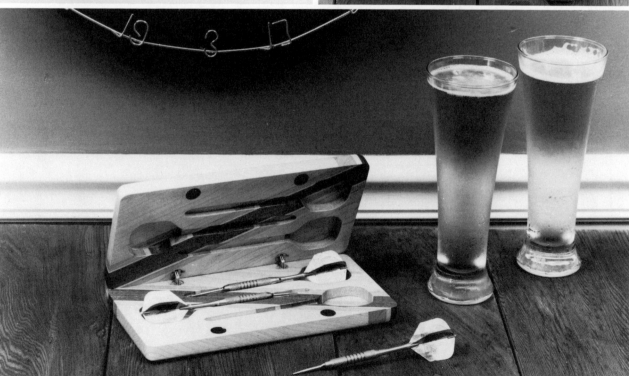

1
Determine the size of the case and the shape of the recesses. Collect the things you want to store in the case. Arrange them on a flat surface in the manner you want to store them in the case. Measure the area — length and width — that the arrangement takes up, and add 1" to each dimension.

This gives you the length and width of the case. Measure the thickness of the thickest item and add ¾". This gives you the thickness of the case.

Place each item on a piece of thick paper or posterboard, and trace around it. Cut out the shapes to make templates for the recesses that you will saw in the case.

2
Select and cut the stock. Since the design for the case is so simple — a single rectangular board — select a piece that is highly figured or has an especially attractive grain pattern. Wood with such defects as burls, curly grain, bird's-eyes, spalting, and interesting knots will make attractive one-board cases. You might also use an exotic wood, such as cocobolo, zebrawood, or rosewood. Make sure the board is somewhat larger than the dimensions you figured for the case.

If you don't have an interesting piece of wood or one large enough to make a case, glue up a board from scraps. Assemble the scraps in a geometric design to create some visual interest. The dart case shown is laminated from ribbons of multicolored wood that stretch from corner to corner. (See Figure 1.) You might also make a traditional *Herringbone Pattern, Concentric Squares, Pie Sections,* or a pattern of your own design.

Once you have made or selected the stock, cut it about ⅛" longer and wider and ⅜" thicker than you calculate you'll need. This extra margin will allow for saw kerfs and sanding.

1/To make stock with diagonal stripes, first laminate a board much larger than you need, using different-colored scraps. Lay out the shape of the case on the larger board at an angle, then cut out the case on a band saw.

CONCENTRIC SQUARES

HERRINGBONE PATTERN

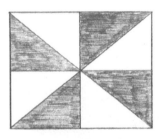

PIE SECTIONS

3 **Cut the bottom and top.** Cut a ¼"-thick slice from both the top and bottom face of the stock, using the band saw to resaw the wood. (See Figure 2.) Sand the sawed surfaces with a belt sander or disk sander to remove the saw marks.

2/Resaw the stock to make the bottom and top of the case. Put a little extra tension on the band saw blade. When sawing through stock of uneven hardness, or through extremely hard stock (such as burls), the blade has a tendency to cup in the cut.

4 **Cut the recesses in the case.** Using your paper templates, trace the shape of the recesses on the *middle* of the stock (not the bottom of the top). Drill holes through the areas where you want these recesses to be *deepest*. On the knife case, the deepest areas are those that hold the handles. On the dart case, they are those that enclose the feathers or "flights." Insert a scroll saw blade through these holes, and cut out the recesses. (See Figure 3.) Save the waste.

Lightly file the sawed edges of the recesses to remove the saw marks. *Don't file away too much stock!* Later, you'll cut plugs from the waste and glue them in the recesses. These plugs won't fit properly if you enlarge the recesses.

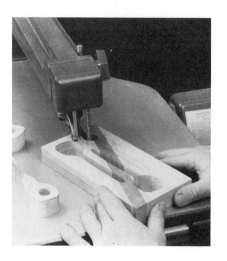

3/When sawing the recesses in the case, use a fine blade with 16 or more teeth per inch. The blade will cut slowly and may break, but it will leave a much smoother surface than a coarser blade.

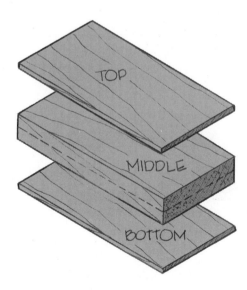

CUTTING THE TOP AND BOTTOM

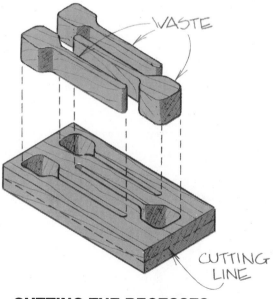

CUTTING THE RECESSES

5 Assemble the case and cut it in half.

Glue the top and bottom to the middle. Carefully align the grain patterns or the stripes as they were before you resawed the wood. After you sand the glue joints clean and flush, you shouldn't be able to tell that the board had ever been cut apart.

Resaw the board again, splitting it in half. (See Figure 4.) Sand the sawed surfaces on a belt sander or disk sander to remove the saw marks. Make sure that the two halves fit back together flush, with no gaps in the seam.

4/After gluing the top and bottom back to the stock, resaw the board in half. The two halves of the case should look like mirror images of each other.

6 Glue plugs in the recesses where needed.

The items that you have to store in a case are rarely of a uniform thickness. For example, the blades of knives are much thinner than the handles. The shafts and tips of darts are much thinner than the flights. For these items to sit properly in the recesses, you must reduce the depth of some areas.

Do this by cutting plugs from the same waste that you sawed from each recess. Sand the saw marks from the cut surfaces of the plugs that will show, then glue them in the recesses. (See Figure 5.) Use epoxy glue to hold the plugs in place. Because of the scroll saw kerf, there will be a tiny gap between each ledge and the sides of the recess. Epoxy glue fills this gap much better than ordinary resin glue. Resin shrinks when it dries; epoxy doesn't.

5/Be very careful to line up the grain pattern or the wood stripes of the plugs and the case. The completed case should look like a single board, both inside and out.

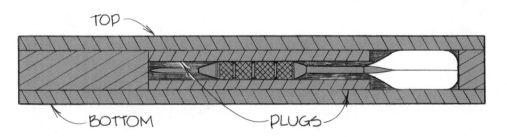

TYPICAL SECTION

7 Hinge the halves of the case together.

To hold the case together, you may use any hinge or catch that you wish. The knife case is fitted with ordinary butt hinges and a clasp — you can see the hardware from the outside. The dart case has concealed "no-mortise" hinges and an inlaid magnetic clasp so the hardware is invisible when the case is closed.

When you have decided what sort of hardware to use, mortise the case for the hinges and the clasp, if necessary. Install the hardware, then sand or file the ends and edges of the case so they are perfectly flush. (See Figure 6.) Round or soften the corners if you want to do so.

6/Sand or file the assembled case to make the ends and edges perfectly flush. A soft-sided or pneumatic drum sander makes it easy to round the hard corners.

8 Finish the case.

Finish sand the outside of the case while it's still assembled. Then remove the hardware and finish sand the inside. Apply a finish that penetrates the wood, such as tung oil or Danish oil. Varnishes and other finishes that build up on the surface will quickly become scratched with use. Built-up finishes may also cause the case to stick together on humid days. After the finish dries, reassemble the case.

> **TRY THIS!** If you want to add an interesting finishing touch to the inside of the case, "flock" the recesses — coat the surfaces with bits of ground-up, brightly colored felt. This makes the recesses look as if you've molded a cloth liner to them. Flocking kits are available through some mail-order woodworking suppliers and at most craft or hobby stores.

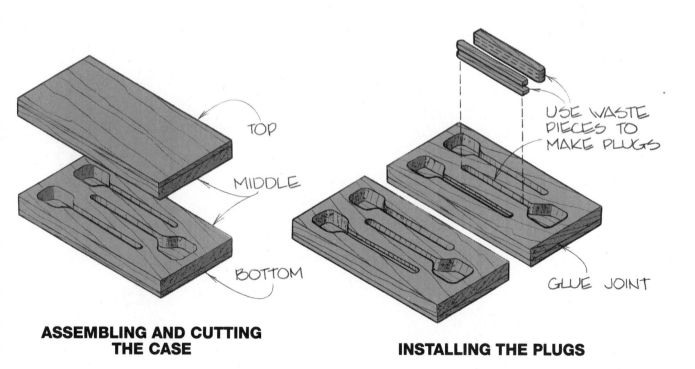

ASSEMBLING AND CUTTING THE CASE

TOP

MIDDLE

BOTTOM

INSTALLING THE PLUGS

USE WASTE PIECES TO MAKE PLUGS

GLUE JOINT

Pigeonholes

Many collectibles and keepsakes are only a few inches tall or wide. They get lost in standard-size shelving units and cabinets. You need a scaled-down cabinet with small shelves to hold these tiny items so you can see them.

These pigeonholes will do the job. The horizontal and vertical boards create many small compartments of various sizes. To display an item in the unit, just choose the pigeonhole that fits it best.

Materials List

FINISHED DIMENSIONS

PARTS

A.	Top/bottom (2)	½" x 2" x 17½"
B.	Sides (2)	½" x 2" x 25"
C.	Short shelves (5)	½" x 2" x 2½"
D.	Medium-short shelves (6)	½" x 2" x 5"
E.	Medium-long shelf (1)	½" x 2" x 7½"
F.	Long shelves (2)	½" x 2" x 10"
G.	Short dividers (3)	½" x 2" x 3½"
H.	Medium-short dividers (2)	½" x 2" x 7"
J.	Medium dividers (4)	½" x 2" x 10½"
K.	Medium-long divider	½" x 2" x 17½"
L.	Long divider	½" x 2" x 24½"

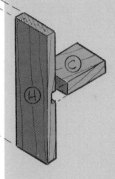

EXPLODED VIEW

HARDWARE

4d Finishing nails (8)
¼" Molly anchors (2)

1 **Select and cut the stock.** To make this project, you need approximately 5 board feet of 4/4 (four-quarters) cabinet-grade wood. The pigeonholes shown are made from cherry, but you can also use pine, oak, walnut, mahogany — almost any species you want, as long as it's straight and clear.

After you choose the stock, plane it to ½" thick. Cut all the pieces to the sizes given in the Materials List. Rip some extra 2"-wide stock and set it aside. You can use this stock to replace parts if you happen to make a mistake in the next few steps.

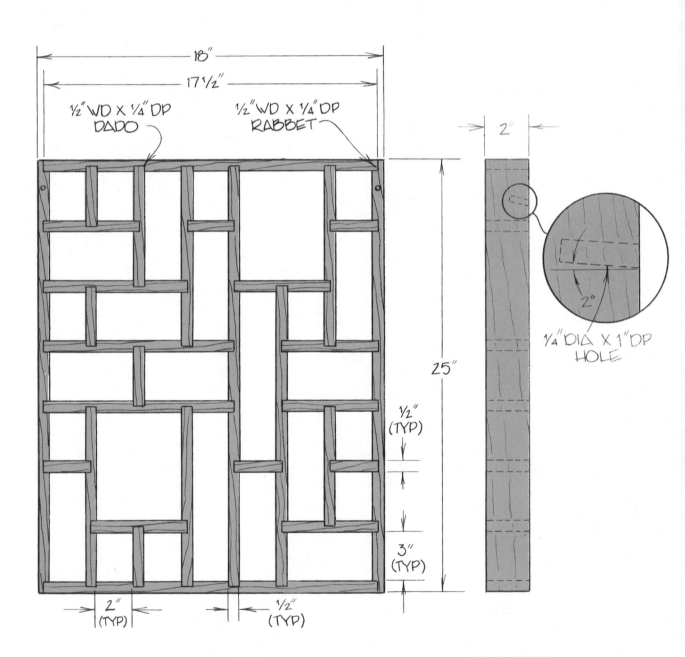

FRONT VIEW   **SIDE VIEW**

2 **Mark the rabbets and dadoes.** The
most important operation in this project is
making the joinery: You *must* position and cut the
rabbets and dadoes as accurately as possible. To help
do this, make a *storystick* to mark the joints. Select a
long, thin piece of scrap or posterboard, and mark it
as shown in the *Storystick Layout*. Note that one side
of this stick measures the shelves, while the other
measures the dividers.

1/Mark the rabbets and dadoes on the faces of each part, using the story-stick, then transfer these marks to the front and back edges with a square. Shade each joint in pencil, as shown, so it's easy to see.

Lay out all the parts on your workbench in the
approximate position in which they will be assembled.
Keep them in this arrangement until you actually assemble the pigeonholes. After you work on one part, put it
back in position *before* you work on another. This will
help you keep track of all the parts and make it easier
to catch mistakes before they happen.

Using the storystick, carefully measure and mark the
joinery on each part. (See Figure 1.) Mark the face, front

edge, and back edge of the part. When all the parts
have been marked, inspect the arrangement on your
workbench. Make sure all the rabbets and dadoes are
indicated. Also make sure you haven't accidentally
marked a rabbet or dado that doesn't need to be cut.

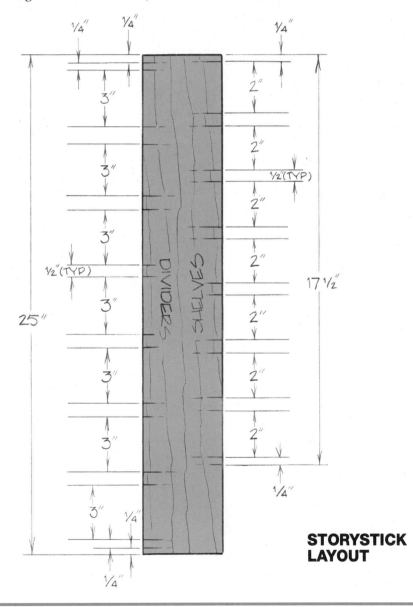

STORYSTICK LAYOUT

3 **Cut the rabbets and dadoes.** Set up a dado cutter or a table-mounted router to cut ½"-wide, ¼"-deep joints. Cut several test dadoes in scraps to check the setup. Remember, this joinery has to be accurate for the project to fit together properly. The test dadoes should be exactly ¼" deep, and the ends of the parts should fit them snugly.

When you're satisfied with the setup, cut the rabbets and dadoes in each part. To help position each joint precisely, make reference marks on the worktable, as shown in Figures 2 through 5.

TRY THIS! To prevent the dado blade or bit from splintering or tearing the wood when it exits the stock, back it up with a scrap.

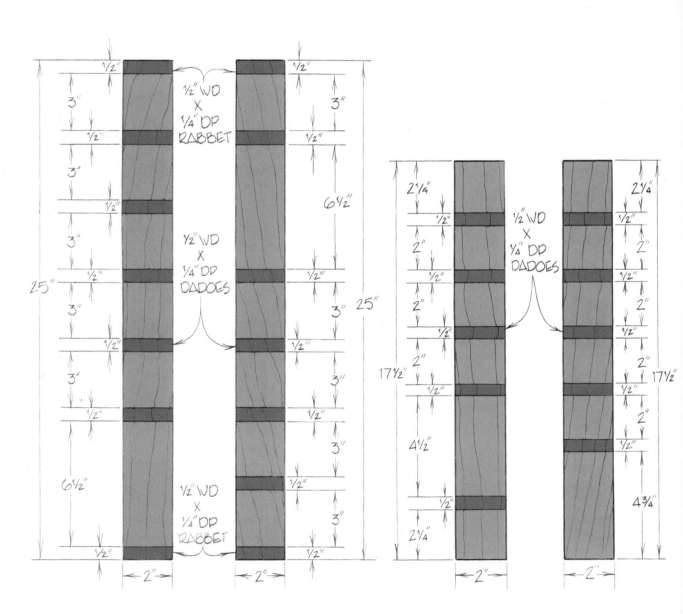

SIDE LAYOUT   **TOP/BOTTOM LAYOUT**

2/When cutting dadoes on a table saw or router table, first cut a test dado in a scrap. After making the joint, pull the stock back over the blade, keeping it firmly in place against the miter gauge. With a pencil, mark the sides of the dado on the worktable. (You can erase the pencil marks later.)

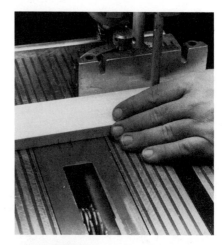

3/Place a part on the worktable, lining up the dado marks with the marks on the worktable. The **back** edge of the part should face the blade. This way, if you cut the first little bit of the dado in the wrong position, you can shift the stock and cut it again. The mistake won't be noticeable in the assembled project.

4/If you're using a dado cutter, double-check the position of the part. Advance it until the edge just touches the cutter. The teeth on either side of the cutter should touch the marks that indicate the sides of the dado.

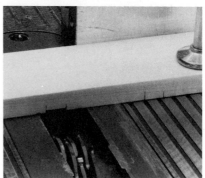

5/Turn on the power tool and carefully feed the part until the cutter begins to cut the wood. Stop and check your work. If the machine is cutting the dado in the proper position, continue. If not, turn the machine off and reposition the part.

4 Assemble the pigeonholes.

Assemble the pigeonholes. Dry assemble the project on the workbench (with the parts resting on their back edges) to test the fit of the joints. When you're satisfied everything fits properly, finish sand all parts. Be careful not to round any adjoining surfaces. Also be careful not to remove too much stock when you sand. If you do, the parts will fit sloppily in the dadoes and rabbets.

Start assembling the project by attaching the top, bottom, and sides. Glue them together, reinforcing the rabbet joints with 4d finishing nails. Set the heads of the nails below the surface. Next, glue the long divider in

place. Then add the remaining parts, working from left to right. Glue just one, two, or (at the most) three parts at a time. Get these positioned just right, then go on to the next part or group of parts. Wipe off any excess glue with a wet rag as you work.

> **TRY THIS!** To keep the parts from slipping after you position them, hold them with masking tape. Remove the tape when the glue has dried.

5 Finish and hang the pigeonholes.

Finish and hang the pigeonholes. Sand all joints clean and flush, and do any necessary touch-up sanding. Then apply a stain or finish. Use a finish that penetrates the wood, such as tung oil or Danish oil. You'll find that building finishes, like shellac or varnish, are much too difficult to rub out. Your hands won't fit inside the little compartments.

Drill a ¼"-wide, 1"-deep hole in the back edge of each side, near the top, as shown in the *Side View*. These

holes should be angled very slightly (just 1˚ or 2˚) toward the top.

Install Molly anchors in the wall where you wish to hang the pigeonholes. Space the anchors *exactly* the same as the holes in the project's sides. Thread the bolts into the anchors, letting about ¾" of each bolt stick out. Put the pigeonholes in place over the anchors, inserting the bolts into the holes.

Museum Table

Museum tables make wonderful display units for small keepsakes. Originally, they were designed to hold small biological specimens, but you can also use them for coins, stamps, shells, rocks, Indian artifacts, and the like. In addition, they make unique side tables or occasional tables.

A typical museum table is just a shallow glass case on four legs. The top of the case opens, letting you reach the objects inside. This top is usually made from thick, tempered glass so you can place heavy objects on it, if you wish. The sides also are glass, to better display the objects in the case. The museum table shown has classic cabriole legs, to blend with a traditional decor. If you prefer contemporary furniture, you may substitute straight or tapered legs.

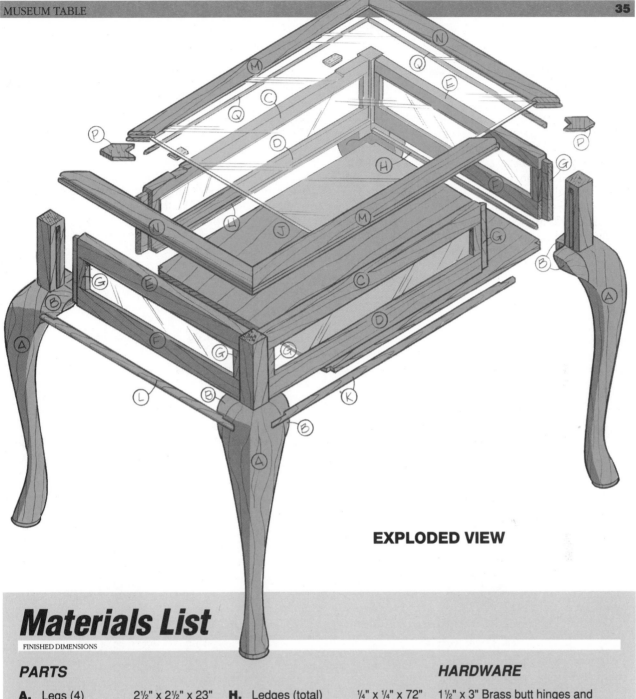

EXPLODED VIEW

Materials List

FINISHED DIMENSIONS

PARTS

A. Legs (4)   2½" x 2½" x 23"

B. Ears (8)   1½" x 1¾" x 2½"

C. Front/back
upper rails (2)   ¾" x 1⅛" x 21⅛"

D. Front/back lower
rails (2)   ¾" x 1⅝" x 21⅛"

E. Side upper
rails (2)   ¾" x 1⅛" x 17⅛"

F. Side lower
rails (2)   ¾" x 1⅝" x 17⅛"

G. Front/back/
side stiles (8)   ¾" x 1⁹⁄₁₆" x 3¼"

H. Ledges (total)   ¼" x ¼" x 72"

J. Bottom   ½" x 16" x 20"

K. Front/back
moldings (2)   ½" x ½" x 22½"

L. Side moldings (2)   ½" x ½" x 18½"

M. Top frame
rails (2)   ¾" x 1½" x 22½"

N. Top frame
stiles (2)   ¾" x 1½" x 18½"

P. Splines (4)   ⅛" x 1" x 2¼"

Q. Glass beading
(total)   ¼" x ⁵⁄₁₆" x 240"

HARDWARE

1½" x 3" Brass butt hinges and
mounting screws (1 pair)

¾" Wire brads (60–72)

¼" x 16⅝" x 20⅝" Tempered glass
pane (for top frame)

⅛" x 3⅛" x 17⅞" Glass panes (2 for
front and back frames)

⅛" x 3⅛" x 13⅞" Glass panes (2 for
side frames)

Felt or leather (16" x 20" – optional)

1

Select and cut the stock. To make this project, you need about 6 board feet of 10/4 (ten-quarters) lumber, 7 board feet of 4/4 lumber, and one quarter sheet (24" x 24") of cabinet-grade ½" plywood. You can use any cabinet-grade wood for this project, but classic furniture is traditionally built from mahogany or, less frequently, cherry or walnut. If you use wood with a prominent, open grain such as oak or ash, consider making the legs straight or tapered. Cabriole legs made from this kind of hardwood often look odd — the complex curves and the swirling grain are hard to focus on. When the legs are cut along simple, straight lines, the grain patterns aren't as visually distracting.

Cut the 10/4 stock into blocks for the legs and ears. Plane 5 board feet of the 4/4 lumber to ¾" thick, and cut all the ¾"-thick parts to the sizes shown in the Materials List, except the top frame members. Cut these parts about 1" longer than specified and don't miter the ends yet. Also cut the plywood bottom slightly larger than needed.

Plane 1 board foot of the 4/4 lumber to ½" thick, and set it aside to make the moldings. Resaw the remaining board foot of 4/4 stock in half, and plane the halves to ¼" thick. Set it aside with the molding stock. Later, you'll use this stock to make the glass beading and ledges. Also, don't cut the splines yet — wait until after you've made the spline grooves.

2

Mortise the leg stock. You must make the mortises in the legs *before* you cut the cabriole shape, while all the surfaces are still flat and the stock is square. You'll find it almost impossible to cut this joinery accurately after you've shaped the legs.

Enlarge the *Leg and Ear Patterns* and trace them on scraps of thin plywood or hardboard. Mark the begin-

ning of the knee — the point on the leg shape where the straight post stops and the curved portion begins. Cut out the shapes with a band saw or scroll saw, and sand the sawed edges. These cutouts will serve as templates to mark the leg stock.

Using a square, mark the knees on all four faces of each leg. Then trace the cabriole shape on the two *inside* faces, carefully lining up the knee marks on the template with those on the leg stock. Lay out the mortises on the inside faces of the post, as shown in the *Corner Joinery Detail*.

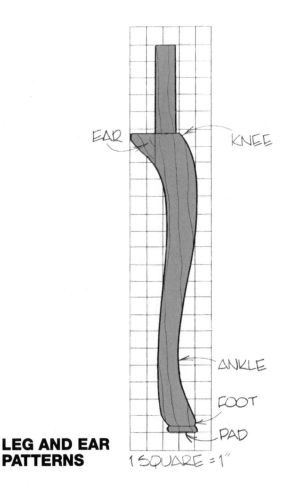

LEG AND EAR PATTERNS

1 SQUARE = 1"

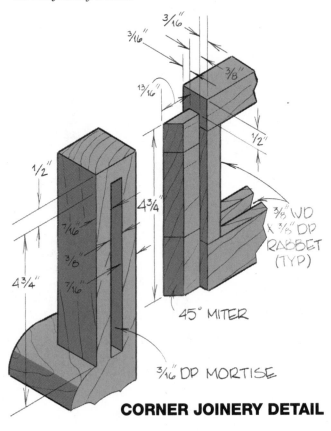

CORNER JOINERY DETAIL

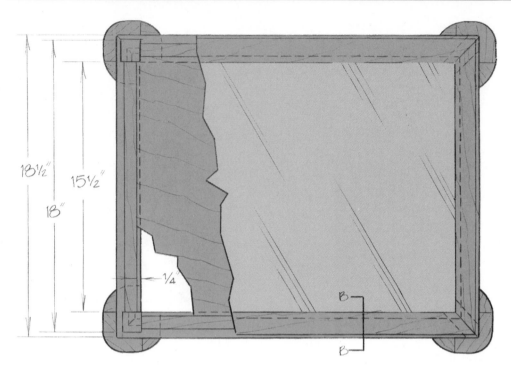

TOP VIEW

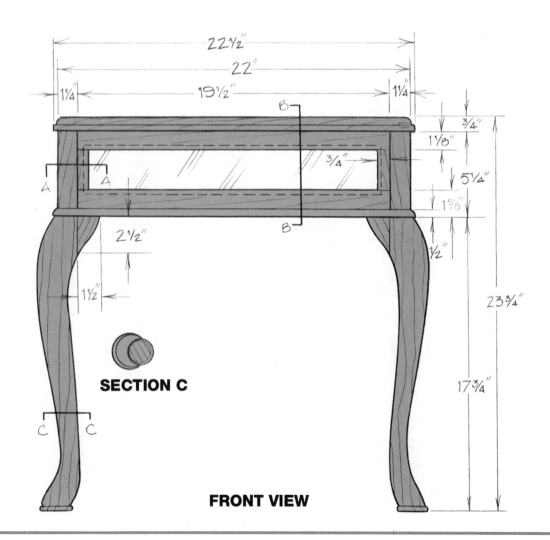

SECTION C

FRONT VIEW

To make each mortise, drill a line of overlapping ⅜"-diameter, ¹³⁄₁₆"-deep holes. (See Figure 1.) Then square up the sides and corners with a chisel. When you finish, the mortise should be ⅜"-wide, ¹³⁄₁₆"-deep,

and 4¾" long. Drill and cut two mortises in each leg. These mortises should intersect inside the post, as shown in *Section A*.

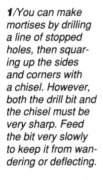

1/You can make mortises by drilling a line of stopped holes, then squaring up the sides and corners with a chisel. However, both the drill bit and the chisel must be very sharp. Feed the bit very slowly to keep it from wandering or deflecting.

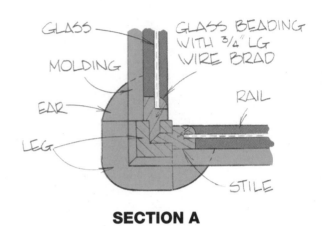

SECTION A

3 Cut and sand the cabriole legs.

Glue the ear stock to the leg stock. (The top end of the ears should be even with the knee marks.) Let the glue dry at least 24 hours, then mark the ear shapes on the two *inside* faces of each ear. Once again, carefully line up the knee marks on the template and the stock. Then saw the rough cabriole shapes by making successive band saw cuts in two adjacent faces of each board. (See Figures 2 through 4.) This technique is called "compound cutting."

Do the final shaping with hand tools. First smooth the posts with a block plane and chisels, taking care to

plane them as square as possible. Then round the legs, feet, and pads with a spokeshave, back saw, files, rasps, and sandpaper. (See Figures 5 through 9.)

> **TRY THIS!** To hold the leg so you can work on it, mount it in a bar clamp. Secure the bar clamp to your workbench with handscrew clamps or bench dogs.

*2/Place the leg on a band saw with one inside face up. Saw one outside edge of the post down to the knee mark, turn the post 90°, and cut the other. Adjust the upper blade guide to clear the ear stock, then cut along the outside curve of one ear. Saw up to the knee mark to free the waste from the post. **Save the waste.***

3/Lower the blade guide so it just clears the leg stock. Cut the inside curve of the other ear and both the outside and inside curves of the leg. Once again, save the waste.

4/Tape the waste back on the stock and turn it 90°. Repeat the cuts you just made: First cut the outside curve of an ear, then the inside curve of the other ear, then the inside and outside curves of the leg. When you remove the tape and the waste, you'll have an S-shaped leg that curves through three dimensions.

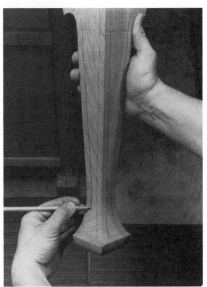

5/To help round the legs evenly and shape them identically, lay out a line parallel to the corners on each face.

6/Mount the leg in a vise, with the foot facing up. Mark a circle on the bottom of the foot to indicate the pad. Using a back saw, cut off the three **outside** corners of the foot. Use the quarter lines as guides, cutting from the left-most line on one face to the right-most line on the adjacent face. Also, saw all four corners from the pad, making an octagon shape.

7/Using the edge of a rectangular double-cut file, round the pad. With a cabinet rasp, round the three outside corners of the foot.

8/Clamp the leg horizontally on your workbench. With the spokeshave and cabinet rasp, chamfer each corner. As before, the chamfers should extend from the left-most quarter line on one face to the right-most line on the adjacent face. To keep from accidentally cutting or nicking the square corners of the post, cover them with masking tape.

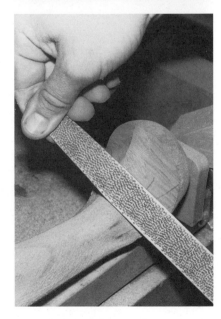

9/After you've made the chamfer, round over the faces of the leg with files and sandpaper, blending them into one another.

4 **Assemble the front, back, and side frames.** Using a dado cutter or a table-mounted router, cut ⅜"-wide, ⅜"-deep rabbets in the inside edges of all the front, back, and side frame parts. Cut the same size rabbet in both outside ends of each stile. The *edge* rabbets are on the *inside* face of both the rails and stiles, while the *end* rabbets are on the *outside* face of the stiles only.

Finish sand the rails and stiles. Glue the frame parts together, lapping the end rabbets on the stiles over the edge rabbets on the rails. Let the glue dry, then sand all joints clean and flush.

5 **Cut the tenons in the front, back, and side frames.** After you have assembled the frames, cut ⅜"-wide, ¹³⁄₁₆"-long tenons in the ends of each frame, as shown in the *Tenon Layout*. Guide the frames with a miter gauge as you make the cuts with a dado cutter or a table-mounted router.

Using a band saw or dovetail saw, cut ½" from the top edge of each tenon. Miter the ends of the tenons at 45° on a table saw so the tenons will fit in the intersecting mortises as shown in *Section A*.

Test fit the tenons in the mortises. If a tenon is too tight, shave a little stock from it or the inside of the mortise. If it's too loose, glue a scrap of veneer to the tenon to build it up. When all the tenons fit properly, dry assemble the frames and legs. Wrap a band clamp around the assembly to keep it together.

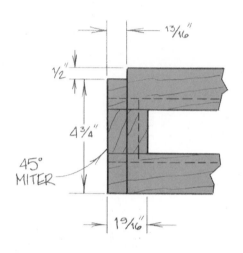

TENON LAYOUT

TRY THIS! To judge where to shave stock off a tenon, rub the inside of the mortise with a soft lead pencil. Slide the tenon in the mortise as far as it will go, then back out again. Inspect the tenon for black marks — these will show where it is too thick or wide.

6 **Cut the top frame splines and grooves.** Measure the assembled case to see if the dimensions have changed from those in the drawing. There will probably be some changes; this is normal for a project that requires so many different cuts and joints. If so, make corresponding changes in the dimensions of the top frame. Then cut the top rails and stiles to length, mitering the ends at 45°.

To join the frame parts, cut a ⅛"-wide, ½"-deep groove in the mitered ends of each rail and stile with a table saw and combination blade. Use a tenoning jig to hold the frame parts at 45° to the saw table while you cut. (See Figure 10.)

Cut ⅛"-thick, 1"-wide splines from scrap stock, making them slightly longer than specified in the Materials List. Also make sure the grain direction runs from *side to side* on each spline. Dry assemble the top frame parts with the splines in place. Mark the shape of the splines, using the inside and outside edges of the frame parts as a straightedge. Cut the shape of the splines on a band saw.

10/When cutting spline grooves in mitered parts, hold the parts in a tenoning jig. Make sure the mitered end rests flat on the work surface.

7

Shape the top frame parts, molding, and glass beading. Using a shaper or a table-mounted router, round over the top inside edges of the top frame rails and stiles, as shown in the *Top Frame Profile*. Round over the top outside edges, and a ⅜"-wide, ½"-deep rabbet on the inside bottom edges.

To make the moldings, round over the edge of a ½"-thick board, at least 3" wide. Then rip a ½"-wide molding from the board. (*Never* try to shape a thin, narrow piece of stock — the wood may come apart in your hands.) Cut it a little longer than you need, and don't miter the ends yet. Repeat until you have made all the moldings. Make the glass beading in the same way, using ¼"-thick stock: Round over the edge of a thick board, then rip the beading free.

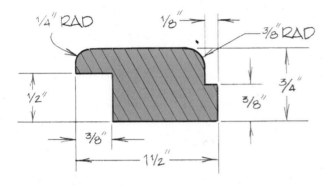

TOP FRAME PROFILE

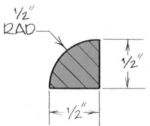

MOLDING PROFILE

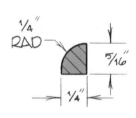

GLASS BEADING PROFILE

8

Assemble the case. Take apart all the assemblies and finish sand the parts you have made so far. Glue the frame tenons in the leg mortises, checking that the case is square as you clamp it together. Wipe away any glue that squeezes out of the mortises with a wet rag.

9

Fit the moldings, ledges, and bottom. Notch the back edge on the moldings to fit against the legs and frames, as shown in the *Molding Layout*. Miter the ends of one molding and clamp it in place. Then miter a second, adjoining molding to fit against the first. Work your way around the case, fitting the moldings to each other. Do any necessary touch-up sanding on the moldings, and glue them in place. The bottom edge of the moldings should be flush with the bottom edge of the frames, as shown in *Section B*.

Rip the ¼"-wide ledges from the remaining ¼"-thick stock, and cut them to fit against the inside of the frames. Glue them in place opposite the moldings, so the bottom edges of the moldings, frames, and ledges are all flush. When the glue dries, sand the bottom edges clean.

Measure the inside of the assembled case and trim the plywood bottom to fit. Notch the corners of the bottom, as shown in the *Bottom Layout*. Glue it in place on the ledges.

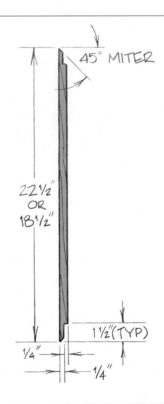

MOLDING LAYOUT

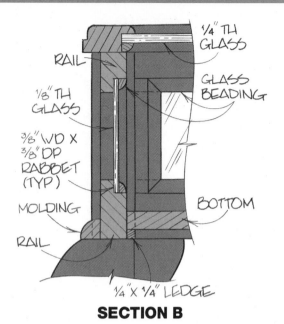

SECTION B

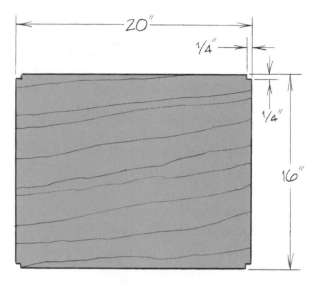

BOTTOM LAYOUT

10 Assemble and install the top frame.
Assemble the parts of the top frame with glue and splines. After the glue dries, sand the miter joints clean and flush.

Mortise the back frame for the butt hinges with a chisel. Put the top frame in place, mark the locations of the butt hinges on its bottom face, and mortise the top frame. Attach hinges to the top frame with screws, then mount the frame on the case.

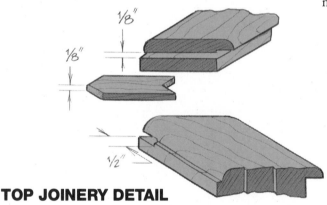

TOP JOINERY DETAIL

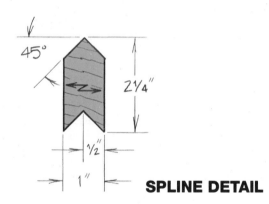

SPLINE DETAIL

11 Finish the completed table.
Remove the top frame from the case and the hinges from the frame. Do any necessary touch-up sanding, then apply a finish to all the wooden parts and assemblies, including the glass beading. Be sure to apply as many coats of finish inside the case as outside. Not only is the inside appearance important, but this will keep the case from distorting.

When the finish dries, place the glass in the frame rabbets. Cut the glass beading to fit inside the rabbets, mitering the corners. Nail the beading in place with wire brads. If the beading tends to split, you can spin the brads into the wood with an electric drill. To do this, mount a brad in the drill chuck. Turn on the drill and drive the brad into the wood, as if you were drilling a small hole. Loosen the chuck and lay the drill aside, then drive the brad the rest of the way with a hammer. Set the head of each brad and, if necessary, touch up the finish on the beading.

Reinstall the top frame on the case. If you wish, cover the bottom with felt, leather, or another suitable material.

Quick-and-Easy Picture Frames

If you're like most people, you *always* have a few pictures around the house that need framing. And if you're like most woodworkers, you've thought that you could make the frames yourself. You may even have con- sidered buying some professional framing equipment — a power miter saw or one of those racy-looking framing clamps.

You're right on one score: You can easily build frames. You don't even need special tools to do a good job. A router, a table saw, and an ordi- nary band clamp are all the equipment you need. Furthermore, by building a simple jig for the saw, you can produce frames of almost any size as quickly and as accurately as you could with the most expensive framing devices.

Note: Because of the many types of frames and infinitely variable picture sizes, there is no Materials List.

1

Design the frame. All of the many frame designs fall into one of three categories:

■ A *face-on* frame is the most common. The decorative shapes are cut into the face — the widest part of the board, which faces out when the frame is assembled. The miter joints are cut across the face.

■ An *edge-on* frame is shaped, mitered, and assembled with the edge facing out, exactly the opposite of a face-on frame. This makes the picture stand out slightly farther from the wall.

■ A *compound-mitered* frame falls somewhere between the first two. The shapes can be cut in the face, the edge, or both. The stock is compound-mitered (cut with both the miter gauge and the blade angled), and assembled so the face of each frame member slopes toward the picture area.

Decide on a category of frame, then design the profile you will cut in the framing stock. Once again, there are three general choices:

■ A *simple* shape can be cut with a single machine setup, using one cutter or bit.

■ A *compound* shape requires two or more setups and may also involve several different cutters or bits.

■ A *laminated* shape is made by gluing together two or more layers of simple shapes. The effect is similar to compound-shaped framing stock.

Each of these shapes has advantages and disadvantages. Simple-shaped framing stock is the easiest and quickest to make, but you cannot make ornate stock without expensive cutters and equipment. By cutting compound shapes, you can make highly decorative stock with common, inexpensive tools, but each additional setup makes precision harder to maintain. You can laminate contrasting woods to create effects that you could never achieve with simple or compound shapes, but laminated shapes also require precision, and you can't make as many profiles as you can by cutting compound shapes.

Consider these advantages and disadvantages in light of your needs, tastes, woodworking experience, and tools. Then design the shapes you want to make, and plan the necessary sequence of cuts.

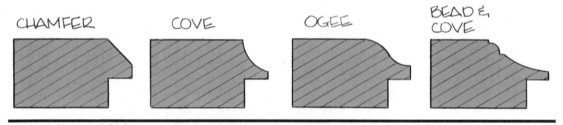

FACE-ON FRAME

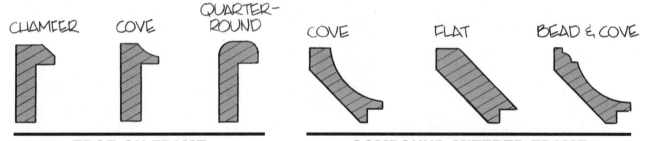

EDGE-ON FRAME   **COMPOUND-MITERED FRAME**

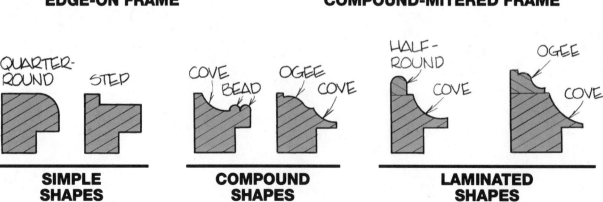

SIMPLE SHAPES   **COMPOUND SHAPES**   **LAMINATED SHAPES**

2 **Select the stock.** You can make framing stock from any wood; however, it should be as clear and as straight as you can find. If the lumber has a bow or a twist, the frame will be out of kilter. It may not be square; there may be gaps at the corners; or it may not lie flat against the wall. Carefully sight down a board before you cut it into framing stock. If it's not straight, discard it and choose another.

3 **Cut the stock to size.** When you have decided on the profile of the framing stock, plane and rip the wood to the thickness and width needed. If any of the shaping cuts will reduce the dimensions, make the rough stock thicker or wider as necessary. Also, cut some extra boards to test your machine setups.

4 **Cut the shapes.** Select the tools, bits, and cutters needed to make the shapes. If you're using a router, mount it in a table. It's possible to shape framing stock with a hand-held router but it's much more time-consuming and more difficult to be precise. You have to clamp and unclamp the stock from your workbench as you cut it.

There are other tools that you can use besides a router. Shapers and molding accessories will also cut frame shapes. (See Figure 1.) It's easiest to cut wide coves on a table saw. (See Figure 2.) You can make chamfers quickly with a jointer. (See Figure 3.) Combine these tools, if necessary, using as many as you need to cut the shapes.

2/To cut a cove on a table saw, pass the wood over the blade at an angle. The angle determines the width and the shape of the cove: A small angle creates a narrow, elliptical cove; a large angle makes a wide, round cove. Use a straightedge to guide the stock, and start with the blade approximately ⅛" above the table surface. Cut the coves in several passes, raising the blade ⅛" with each pass until the cove is as deep as needed.

1/Shapers and molding accessories are easier to set up and more precise than routers. Their design makes them simpler to adjust and align, and their greater mass helps maintain precision over a series of cuts.

3/When cutting chamfers or bevels with a jointer, always tilt the fence **toward** the blade, making an angle of **less** than 90˚ with the table. This prevents the stock from slipping as you joint it. The cut is more accurate and also safer to perform.

No matter what tools you use, always employ a fence or a straightedge to guide the stock, even if you're cutting with a piloted bit. Hold the stock firmly against the fence and the worktable with fingerboards. (See Figure 4.) This will provide extra support and guidance as you cut, freeing you to concentrate on feeding the stock.

Feed the stock at an even speed *against* the rotation of the cutter. (See Figure 5.) You must pass the board across the worktable quickly enough to prevent the bit from burning the wood, but not so quickly that the stock chatters and the tool bogs down. The right speed depends on the tool and the wood. Finding that speed is a matter of practice.

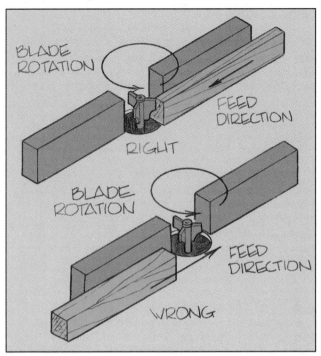

4/For safety and accuracy, use a fence and finger-boards to guide the framing stock as you rout the shapes. The fingerboards help to prevent kickback.

5/Always feed framing stock **against** the rotation of the cutter. If you feed it in the same direction as the rota- tion, the cutter may pull the stock out of your hands and hurl it across the workshop.

---

5 Cut the rabbet in the back of the stock.

After shaping the framing stock, cut a rabbet in the back inside edge (or face). This will hold the glass, mat, picture, and backing in the completed frame. You can make this joint in one step with a dado accessory, or in two steps with an ordinary saw blade.

If you're using a table saw, cut the small dimension of the rabbet first. Then rotate the stock 90° and cut the large dimension. Be careful when ripping rabbets. Position the framing stock and the rip fence so the waste is not caught between the saw blade and the rip fence. (See Figure 6.) If it is, the blade will kick back the waste like a missile as you finish the cut.

Note: When rabbeting framing stock that is designed to be compound-mitered, you must either hold the wood at an angle or tilt the blade as you cut the rabbet. (See Figure 7.) This angle should match the slope of the assembled frame, so the rabbet in the assembled frame will be square to the glass, picture, and backing.

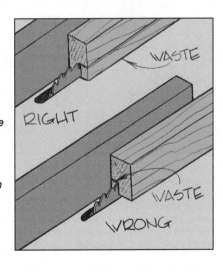

6/Never cut a rabbet so the waste is caught between the blade and the fence — the blade will fling the waste at you as you finish the cut. Make sure the waste ends up on the side of the blade that's **oppo-site** the fence.

7/For compound-mitered framing stock, cut the rab-bets at the same angle as the slope of the frame. Tilt the blade, if you can. If you can't, build a simple jig that attaches to the rip fence to hold the wood at the proper angle.

6 Miter the frames. The most tedious and difficult part of framing is cutting the miters accurately. *Both* the miter angle *and* the length of the frame members must be precise or there will be gaps in the miters. The simple jig shown helps to achieve the necessary precision. It attaches to the miter gauge to support the frame member along its full length as you cut. An adjustable stop automatically gauges the length of a frame member, so you can duplicate other members *exactly*. Furthermore, the sliding scale on this jig can be set to measure the *inside* dimensions of the frame. Simply position the scale to compensate for the dimensions of the framing stock and the rabbet; then measure the dimensions of the picture and set the stop accordingly.

To set up the jig to make the miters, first fasten it to the face of the miter gauge. Then adjust the angle of the blade and/or miter gauge. If you're cutting face-on frames, angle the miter gauge at 45°. If you're cutting edge-on frames, tilt the blade at 45°. For compound-mitered frames, adjust the angles of *both* the miter gauge and the blade according to the following chart:

Slope of Frame	Tilt of Saw Blade	Angle of Miter Gauge
5°	44¾°	85°
10°	44¼°	80¼°
15°	43¼°	75½°
20°	41¾°	71¼°
25°	40°	67°
30°	37¾°	63½°
35°	35½°	60¼°
40°	32½°	57¼°
45°	30°	54¾°
50°	27°	52½°
55°	24°	50¾°
60°	21°	49°

***Note:** All of these angles are textbook settings. The actual settings may vary slightly, depending on how your table saw is aligned and adjusted.*

Note: These plans are drawn for a yardstick that reads from the left. (Most sticks read from the left on one side, and the right on the other.) If using a yardstick that reads from the right only, flop the plans — or turn the yardstick upside down.

TRY THIS! When setting the miter gauge or the saw blade to a simple 45°, use a drafting triangle as a guide. This will ensure the angle is *precisely* 45°.

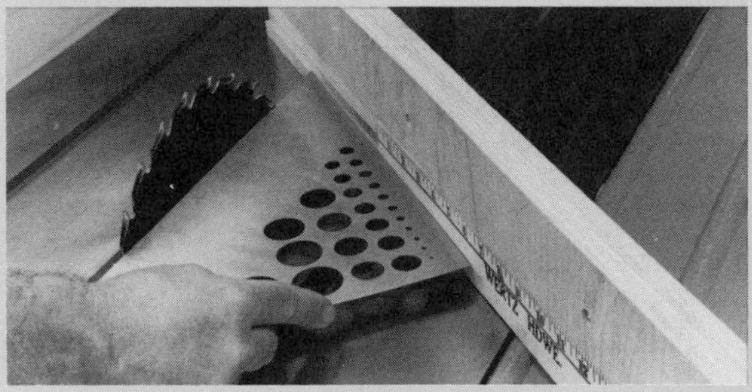

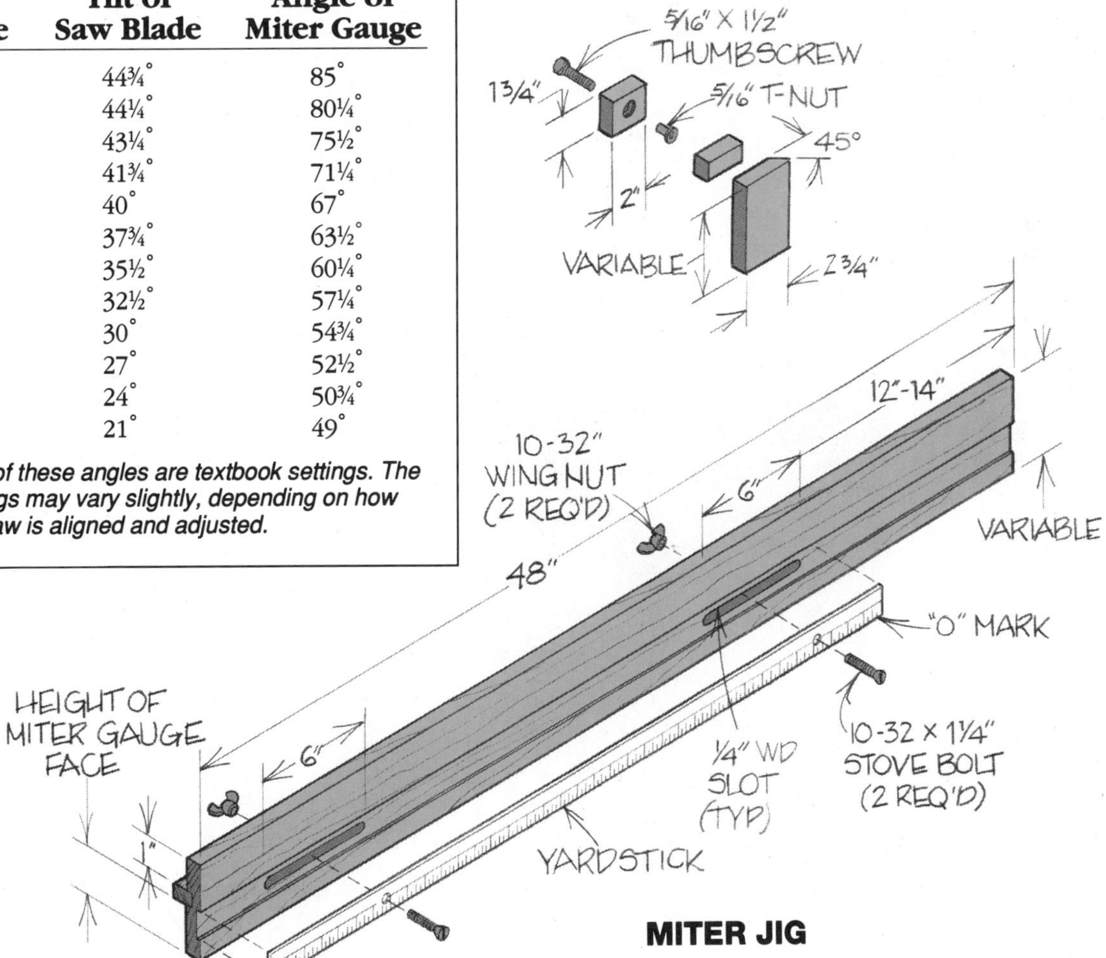
5/16" X 1½"
THUMBSCREW
1¾"
5/16" T-NUT
45°
2"
VARIABLE
2¾"
12"-14"
10-32"
WING NUT
(2 REQ'D)
6"
VARIABLE
48"
"0" MARK
HEIGHT OF
MITER GAUGE
FACE
6"
1"
¼" WD
SLOT
(TYP)
10-32 X 1¼"
STOVE BOLT
(2 REQ'D)
YARDSTICK

MITER JIG

Make several test cuts to determine if the angle is precise. Set the stop about 4" away from the blade, then cut four short pieces of framing stock. Put them together and inspect the miter joints. If the miters gap toward the *outside,* the angle is *less* than 45° and must be increased. If they gap on the *inside,* it's *more* than 45° and must be decreased. The angles for compound miters can be adjusted in a similar way: Outside gaps show the miter angle is too small; inside gaps show it's too large.

Once you've adjusted the saw, set the sliding scale on the jig. The "0" mark on the scale should be the proper distance away from the saw blade. To determine this distance, subtract the rabbet width from the framing stock width, and multiply by 2:

$$Distance = (Frame\ width - rabbet\ width) \times 2$$

If you're making a face-on frame, measure this distance from the "0" to the point of the V formed by the saw blade and the framing jig. If you're making an edge-on or compound-mitered frame, measure it from "0" to the point of the V formed by the saw blade and the worktable. (See Figure 8.)

Measure the width and height of the picture you want to frame. If you wish, add $\frac{1}{16}$" to $\frac{1}{8}$" to both dimensions so the picture will fit loosely in the rabbets of the assembled frame. Set the stop on the jig to the first dimension that you figured and cut two frame members. Then reset the stop to the second dimension and cut two more. (See Figure 9.)

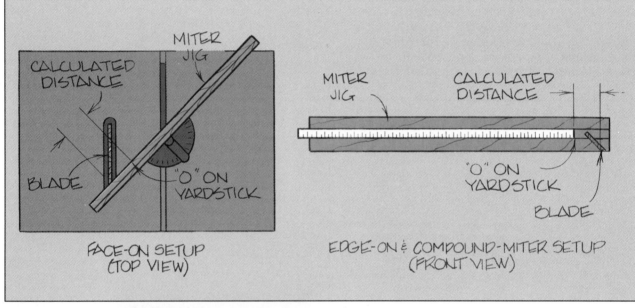

FACE-ON SETUP (TOP VIEW)

EDGE-ON & COMPOUND-MITER SETUP (FRONT VIEW)

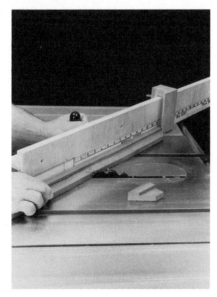

8/The sliding scale must be positioned differently, depending on the sort of frame you're making.

9/To cut a frame member, first miter one end of the stock. Turn the stock end for end and butt the mitered end against the stop. Then cut a second miter.

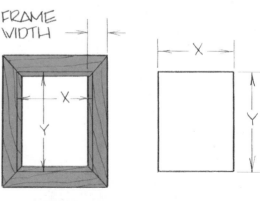

MEASURING DIAGRAM

TRY THIS! The miter joints of picture frames usually don't have to support much weight, so they rarely require reinforcement beyond the brads mentioned above. However, if you need to make a frame that must withstand a heavy load, one of the best ways to strengthen the miter joints is with *splines*. On a table saw, cut a ⅛"-wide groove in the mitered ends of each frame member. Use a tenoning jig to hold the miter flat on the saw table. Then glue the frame together with ⅛"-wide splines in the grooves. The grain direction of each spline must *cross* the miter, so it holds the adjoining pieces together.

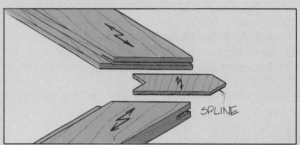

SPLINE

7 **Assemble the frames.** Finish sand the frame members, then dry assemble them to check the fit of the frame. The miters should go together with no gaps. When you're satisfied with the fit, spread glue on the mitered ends. Use a little more glue than you normally would, because the end grain will soak it up. Clamp the members together with a band clamp, and wipe away any glue that squeezes out of the miter joints with a wet rag.

Check that the frame is square. If you've cut the miters accurately, the frame should be self-squaring. That is, the members will automatically pull themselves together at right angles when you tighten the band clamp. Use the metal corners provided with each band clamp to help align the frame members at the corners. (See Figure 10.)

10/To help align the corners of the frame, use the metal corners that come with each band clamp. To keep these metal fixtures from marring the framing stock, place a thin piece of leather or rubber between the metal and the wood.

TRY THIS! Depending on the slope of the frame, you may not be able to use a single band clamp to join compound-mitered frame members. Instead, use *four* band clamps, wrapping them around the frame so they cross each other, as shown.

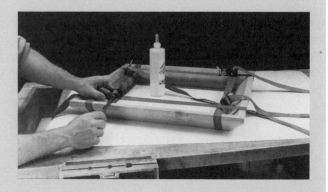

Let the glue cure for at least 24 hours, so the miter joints attain maximum strength. Reinforce the joints by driving several wire brads through each corner. Drive these brads from *both* directions, so they cross at 90° in the wood as shown. This will prevent the corners from pulling apart. Set the heads of the brads below the surface of the wood.

CORNER JOINERY DETAIL

8

Apply a finish to the frame. Cover the brad heads with wood putty or stick shellac, and do any necessary touch-up sanding on the completed frame. Apply a finish to the frame, coating both the front and the back surfaces of the wood.

9

Mount the picture in the frame. Cut a piece of cardboard or posterboard to use as a backing for the picture, and cut a pane of glass to fit the frame. If you want, cut a mat board to mount the picture. (See the section on making a Mat-Cutting Jig for Picture Frames.) Clean the glass thoroughly, then sandwich the pieces — glass, mat, picture, and backing — and place them inside the frame. Hold them in the frame with wire brads, tacked into the sides of the rabbets.

> *TRY THIS!* Many glass shops sell special non-reflective glass for picture frames. This material helps eliminate the glare from windows and nearby lights that prevents you from seeing a picture clearly.

If you've made a small frame, install an eye screw in the back surface of both *side* frame members, and string picture-hanging wire between the two screws. If you've made a large frame, install an eye screw in the back surfaces of all four frame members — top, bottom, and sides. String the wire from the bottom screw to the top, then to one side, then to the other side, as shown in the *Picture-Hanging Wire Installation Detail*. Tie off the wire only at the bottom and last side screw — just run it

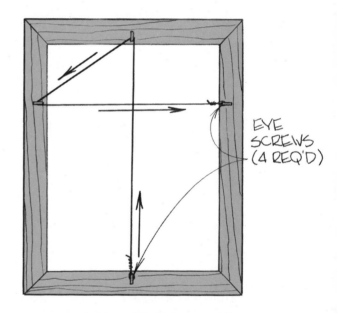

PICTURE-HANGING WIRE INSTALLATION DETAIL

through the other two screws without tying or wrapping it. This arrangement helps hold the frame together and keeps the top, bottom, and sides from bowing. As the weight of the frame stretches the wire tight, it draws the frame members in snugly against the edge of the glass.

Mat-Cutting Jig for Picture Frames

Often, photographs and pictures are *matted* before they're framed. The framer cuts a hole in textured posterboard or *mat board*, and the picture is placed behind this board so it shows through the hole. The mat creates a border around the picture — a frame within the frame, so to speak. Properly done, a mat draws your eye into the picture and helps enhance its beauty.

Cutting your own mat is a simple job, provided you have a mat knife. This tool looks like a block plane and holds a razor-sharp blade. (See Figure A.) Usually the blade is mounted at an angle so that it makes a bevel as it cuts the mat board. You can buy these knives at the same place you buy mat board — art supply stores or stores that sell picture framing materials.

You'll also need a mat-cutting jig to guide the knife. There are many commercial jigs available, but you can easily make your own. Cut a straightedge from a hardwood such as maple or birch, and hinge it to a cutting surface. The cutting surface should be fairly soft — cabinet-grade fir plywood or untempered hardboard works best. Avoid hardwood plywood, tempered hardboard, and particleboard.

To use the jig, position the mat board under the straightedge. Clamp the unhinged end of the straightedge to the cutting surface — this will hold the board in place. Place the knife against the straightedge and push the point of the blade into the mat. Then push the knife along the straightedge, cutting the mat. (See Figure B.) Release the straightedge and reposition the mat for another cut. Repeat, until you've cut all the sides of the border.

A/Mat knives cut the mat board on a bevel.

B/The straightedge guides the mat knife and holds the mat board in place.

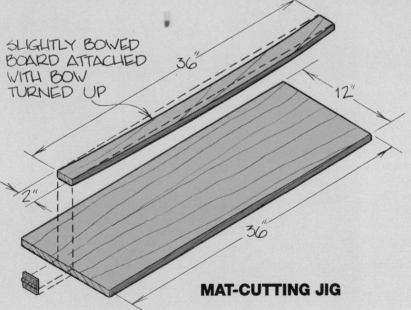

SLIGHTLY BOWED BOARD ATTACHED WITH BOW TURNED UP

36"

12"

2"

36"

MAT-CUTTING JIG

Watch Case

Not all collections require large, elaborate display cases. Some consist of just a few precious objects that can be displayed in something small and simple.

Most antique watch collections, for example, would be lost in a large cabinet. Old pocket watches in working condition are rare and expensive, so most collectors have only a few to display. The small watch case shown is just a shallow box with a glass door. The watches hang from pegs on the back panel. It's extremely simple, but it makes an attractive display.

You can easily adapt this design to other small collections. Add a few more pegs and use it to organize your keys. Install shelves to show off miniatures. Change the width, height, or depth of the case to display bells, figurines, toys, and dozens of other small items. If you substitute a mirror for the back panel, you can see both sides of your treasures. ✳

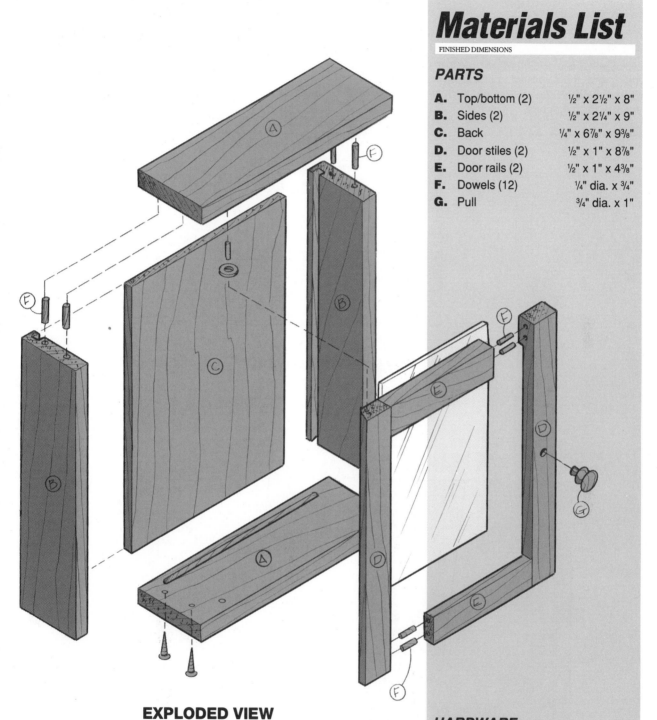

EXPLODED VIEW

Materials List

FINISHED DIMENSIONS

PARTS

A.	Top/bottom (2)	½" x 2½" x 8"
B.	Sides (2)	½" x 2¼" x 9"
C.	Back	¼" x 6⅞" x 9⅜"
D.	Door stiles (2)	½" x 1" x 8⅞"
E.	Door rails (2)	½" x 1" x 4⅜"
F.	Dowels (12)	¼" dia. x ¾"
G.	Pull	¾" dia. x 1"

HARDWARE

#8 x 1¼" Flathead wood screws (4)
¼" dia. x ¾" Metal pins (2)
¼" Flat washers (2)
Glass pane (⅛" x 4¾" x 7¼")
Glazing points (4)
Small screw hooks (3–4)

1

Select and cut the stock.   You can build this project from almost any wood. The watch case shown is made of white pine and painted with milk paint. If you prefer a natural wood finish, use a species with an attractive grain, such as maple, cherry, walnut, or mahogany.

You'll need about 2 board feet of 4/4 (four-quarters) lumber, and a scrap of cabinet-grade ¼" plywood. Cut a small piece of the rough lumber to make a turning block for the door pull, then plane the rest ½" thick. Cut all the parts to the sizes in the Materials List (except the door pull).

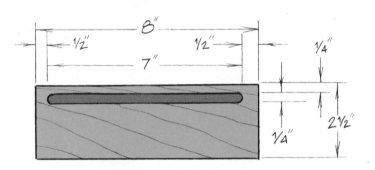

TOP/BOTTOM LAYOUT

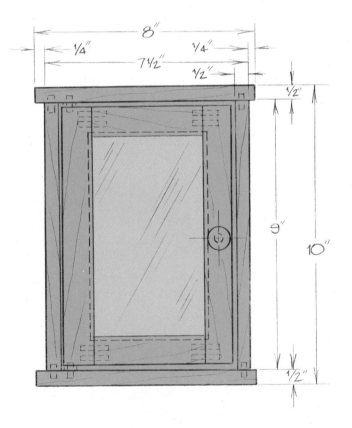

FRONT VIEW

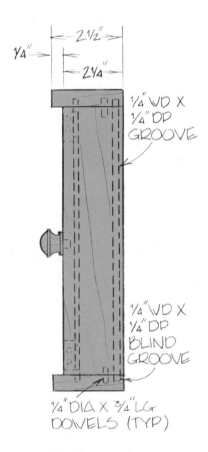

SIDE VIEW

2 Cut grooves in the top, bottom, and back.

Using a router and a straight bit, cut ¼"-wide, ¼"-deep grooves in the sides, top, and bottom, as shown in the *Side View* and *Top/Bottom Layout*. The groove in each side runs the entire length of the part, but those in the top and bottom are *blind* at both ends. Stop these blind grooves ½" from the ends of the boards.

> **TRY THIS!** Before you rout a blind groove, drill holes to mark the beginning and end. (The holes should be the same diameter and depth as the width and depth of the groove.) These holes will show you where to start and stop routing.

3 Assemble the door frame.

Lay out the door parts and mark the locations of the dowels. Using a doweling jig, drill a ¼"-diameter, ⅜"-deep dowel hole at each mark. Also drill a ¼"-diameter, ¼"-deep hole in one door stile for the door pull. Assemble the rails and stiles with dowels and glue.

Let the glue dry for at least 24 hours. Sand the dowel joints clean and flush, then clamp the door frame to your workbench. Using a piloted rabbeting bit, rout ¼"-wide, ¼"-deep rabbets around the *inside edge* of the door. (These rabbets will hold the glass pane.) Square the corners of the rabbets with a hand chisel.

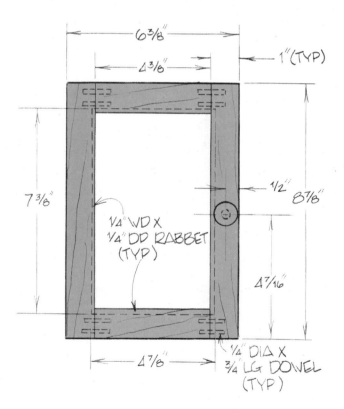

DOOR DETAIL

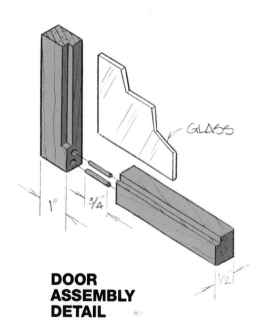

DOOR ASSEMBLY DETAIL

CASE
EXPLODED VIEW

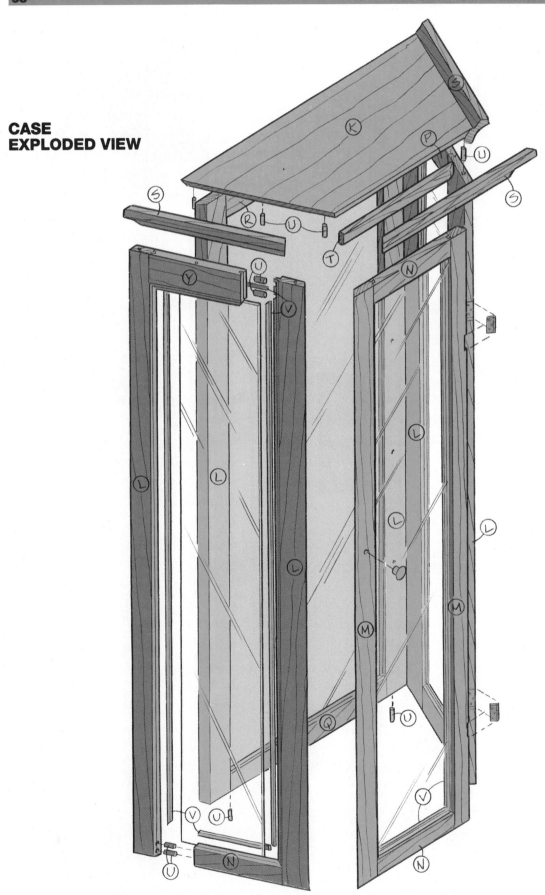

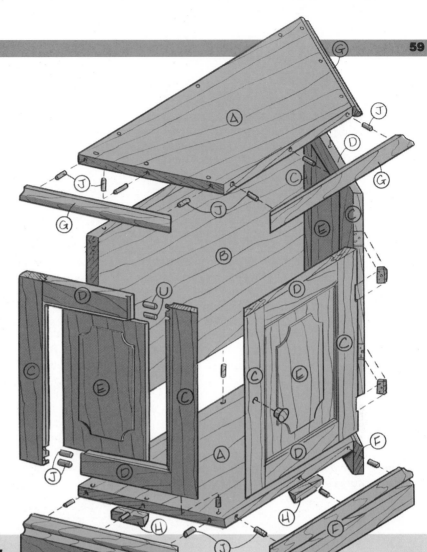

**BASE
EXPLODED VIEW**

Materials List

FINISHED DIMENSIONS

PARTS

Base

A.	Base top/bottom	¾" x 10¼" x 23⅞"
B.	Base back	¾" x 16" x 22⅛"
C.	Base stiles (6)	¾" x 2" x 16"
D.	Base rails (6)	¾" x 2" x 8⅜" *
E.	Base panels (3)	⅝" x 8¼" x 12⁷⁄₁₆"
F.	Base moldings (3)	¾" x 3" x 12¾"
G.	Waist moldings (3)	¾" x 1" x 13"
H.	Glue blocks (3)	¾" x ¾" x 4"
J.	Dowels (30)	⅜" dia. x 2"

Case

K.	Case top	¾" x 11" x 25⅝"
L.	Case stiles (6)	¾" x 2" x 52¼"
M.	Case door stiles (2)	¾" x 2" x 51½"

N.	Case side bottom rails and door rails (4)	¾" x 2" x 8⅜" *
P.	Case side top rails (2)	¾" x 2¾" x 8⅜" *
Q.	Case bottom back rail	¾" x 2" x 18⅝" *
R.	Case top back rail	¾" x 2¾" x 18⅝" *
S.	Case moldings (3)	¾" x 2⅞" x 13⅞"
T.	Front top spacer	¾" x 1" x 12⅛"
U.	Dowels (74)	⅜" dia. x 2"
V.	Glass beading (total)	¼" x ⁹⁄₃₂" x 540"

These measurements may change slightly, depending on the coping cutter used to join the frames.

HARDWARE

#8 x 1¼" Flathead wood screws (20)
#8 x 1½" Flathead wood screws (28)
#6 x 1" Flathead wood screws (9)
#5 x ½" Flathead wood screws (12)
¼" Flat washers (12)
⅝" Wire brads (100)
⅛" x 18½" x 47⅞" Mirror
⅛" x 8¼" x 47⅞" Glass panes (3)
¼" x 8⅝" x 22" Glass shelves (2–3)
Shelving support pins (8–12)
2½" x 1¾" Brass hinges and mounting screws (2 pairs)
1" Brass pulls (2)
Magnetic catches and mounting screws (3)

1

Select and cut the stock. To make this curio cabinet, you'll need approximately 18 board feet of 4/4 (four-quarters) hardwood lumber and one 4' x 8' sheet of ¾" cabinet-grade plywood. The plywood veneer should match the hardwood. The cabinet shown was built from walnut lumber and walnut-veneered plywood. You can also use mahogany, red oak, birch, white pine, or cherry.

When purchasing the hardwood, select the straightest stock you can find. Avoid stock with burls, knots, or other defects, no matter how striking the grain pattern. *This is extremely important!* Straight grain serves best for this project. If the rails and stiles of either the case or base frames are even slightly crooked, they will distort the cabinet. This will prevent the doors from fitting properly.

Before you begin cutting the wood you have chosen, decide how you want to join the frames. As designed, the rails and stiles are joined with "coped" joints. This requires special equipment that you may not have or want to invest in. If so, you can leave the frames plain and join the members with dowel joints. However, you'll need to adjust the dimensions of the rails and stiles on the Materials List.

When you have reviewed the plans and made any necessary changes, plane 3 board feet of lumber to ¼"

thick, another 3 board feet to ⅝", and the rest to ¾". Cut the parts *precisely* to the sizes given in the Materials List, with these exceptions:

■ Rip all the stiles ⅛" to ¼" wider than needed.

■ Cut the base and top moldings 1" longer than needed.

■ Do not cut the waist moldings and the glass beading yet.

Make the case top and the base top, bottom, and back from plywood. Use ⅝"-thick stock for the panels and ¾" stock for the rest of the parts. (Set the ¼" stock aside to make the glass beading.) Cut the case top, base top, and base bottom to trapezoid shapes, as shown in the *Base Top/Bottom Layout* and *Case Top Layout*. Bevel the front and side edges of the case top at 45°, as shown, but do not cut any other bevels.

Note: By cutting the rails and stiles to the precise lengths needed, you'll save some work later on. However, you'll have to be extra careful when you mill the joints — you have *no* margin for error. To help prevent mishaps, make several extra rails and stiles from scrap wood. Use these to check your machine setups thoroughly *before* you cut good lumber. You might also make one or two extra parts from good hardwood, just in case.

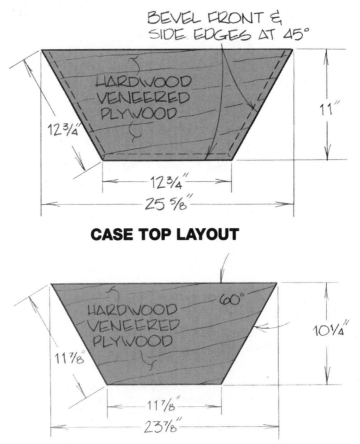

CASE TOP LAYOUT

BASE TOP/BOTTOM LAYOUT

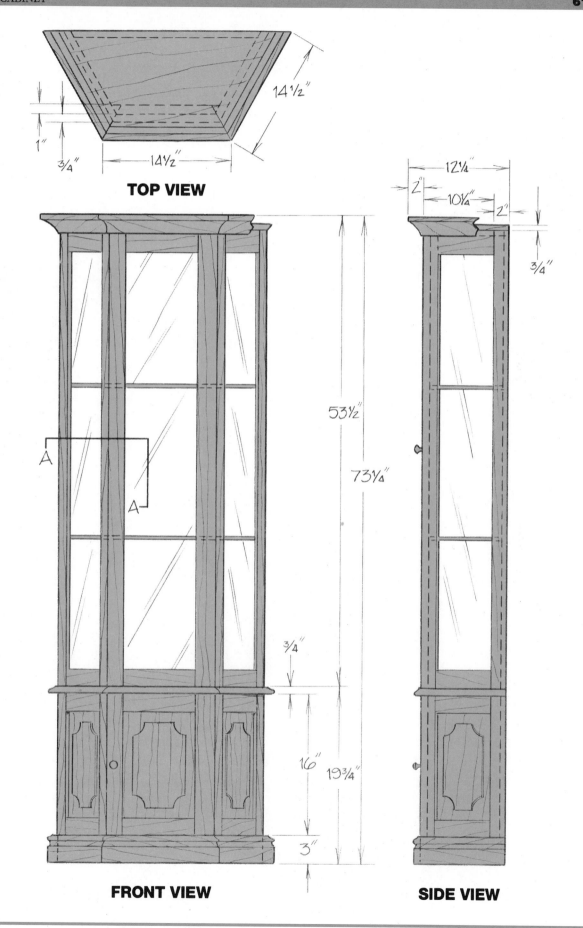

TOP VIEW

14½"

14½"

1"

¾"

FRONT VIEW

53½"

73¼"

¾"

16"

19¾"

3"

A

A

SIDE VIEW

12¼"

2"

10¼"

2"

¾"

2 Cut the rabbets in the rails and stiles.

Use a dado cutter or a router to cut a ¼"-wide, ⅜"-deep rabbet in the *inside back* edges of all the rails and stiles. Carefully check the machine setup before you cut; any error will affect the fit of the frame members.

Note: Some cutter sets make the rabbet and shape the edge of a board at the same time.

3 Cut the coped joints in the frames.

As mentioned earlier, all the frames in this project are joined by coped joints, reinforced with dowels. A coped joint is a molded or shaped joint. The ends of the rails fit over the shaped edge of the stiles, as shown in the *Door Joinery Detail*. To make these joints, you need a special set of matched cabinet door cutters (negative and positive, or "coping" and "sticking") for a shaper or molder.

Make the sticking part of the joint first — the part that sticks into the coped part. Mount the sticking cutter on the shaper or molder, and set up the machine to cut a shaped tongue, ¼" wide and ⅜" from the back surface of the wood. Test the setup with scraps, then shape the inside edges of all the rails and stiles. (See Figure 1.) Use the shaper fence or rip fence to guide the stock.

Change cutters to make the *coped* portion of the joint. Carefully set up the machine so it will cut a matching profile in the end of the rail stock *without* shortening the stock. Cut several test pieces, checking their fit with the edges of the stiles. When the wood fits properly, cut both ends of each rail. (See Figure 2.) If you're cutting this joint on a shaper, use a miter gauge to guide the stock. If you're working with a molder, guide the stock with a tenoning jig.

*1/To make a coped joint, you need a special set of cabinet door cutters. First, make the **sticking** portion of the joint in the edges of both the rails and stiles.*

*2/Then cut the **coped** portion in the ends of the rails only. Since you're cutting end grain, back up the wood with a piece of scrap as you cut it. This will prevent any splintering where the cutter exits the stock.*

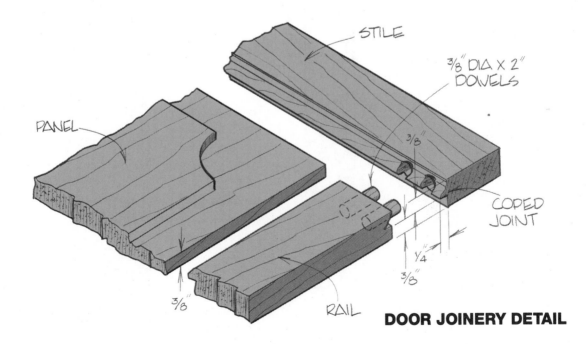

DOOR JOINERY DETAIL

TRY THIS! Some professional cabinet-makers prefer to make coped joints in *three* steps instead of two. First, they cut the sticking portion in the test pieces (but not good lumber). Then they cut the coped portions, fitting the joint to the test pieces. Finally, they cut the sticking portion again, this time in good lumber, and fit it to the coped parts. This extra step helps them catch problems with the setup or the machinery before they waste good lumber.

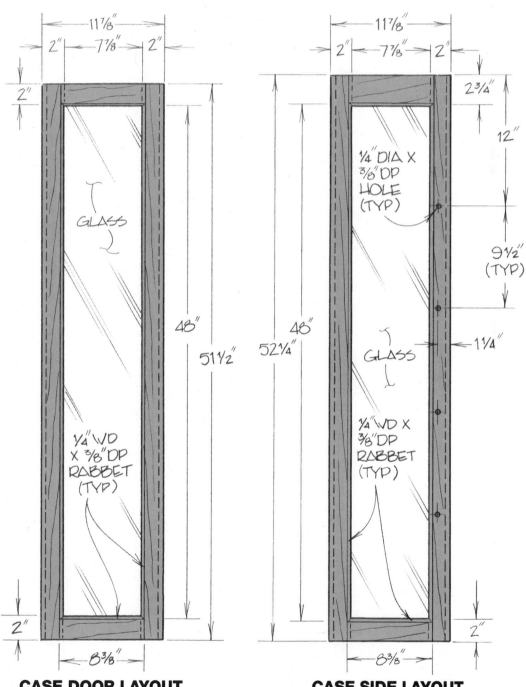

CASE DOOR LAYOUT   **CASE SIDE LAYOUT**

4

Drill holes for reinforcing dowels and shelving supports. Coped joints are not very strong; they must be reinforced if they are to hold. The easiest way to reinforce them is with dowels.

Temporarily dry assemble the frames, and mark the positions of the dowels on both the rails and the stiles. Take the frames apart and drill ⅜"-diameter, 1"-deep dowel holes at the marks. Use a doweling jig to align the holes.

Also drill ¼"-diameter, ⅜"-deep holes in the *inside* surface of the case back stiles and the case side front stiles, as shown in the *Case Back Layout* and *Case Side Layout*. These holes will hold shelving supports when the case is assembled.

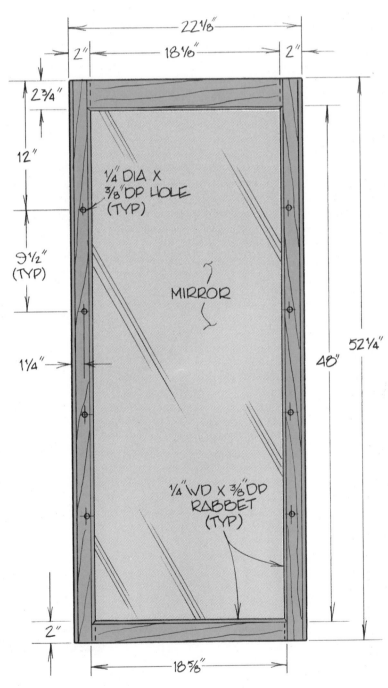

CASE BACK LAYOUT

5

Assemble the frames. Finish sand all the rails and stiles, then assemble the frames with dowels and glue. Use clamps, and keep each frame square, with the stile ends and the outside rail edges flush. After the glue dries, sand each joint clean.

6

Bevel the edges of the frames and base back. Bevel the *outside* of the *side* edges at 60°. Also bevel the edges of the plywood base back while you're set up to cut the case back frame. Remember, the bevels on the case and base backs must face toward the *front;* all the other bevels face the *back*.

7

Assemble the case. Finish sand the case top, base top, base bottom, and front top spacer. Also do any necessary touch-up sanding on the frames. Then assemble the parts of the case in this order:

■ Attach the case back frame to the case side frames with glue and flathead screws. Countersink the screws, as shown in *Section A*.

■ Attach the base back to the base side frames in the same manner.

■ Place the base bottom on the floor of your shop. Temporarily stack the base assembly, base top, case assembly, and case on it. Carefully align all the parts and assemblies. Mark the parts for dowel holes, then take them apart and drill them. Use a drill press to align the holes in the base bottom, base top, and case top, and a doweling jig to align them in the other parts.

■ Reassemble the parts of the base and case with dowels and glue. Wipe away any excess glue with a wet rag.

■ Carefully fit the front top spacer to the case, notching the ends with a band saw. Glue it in place.

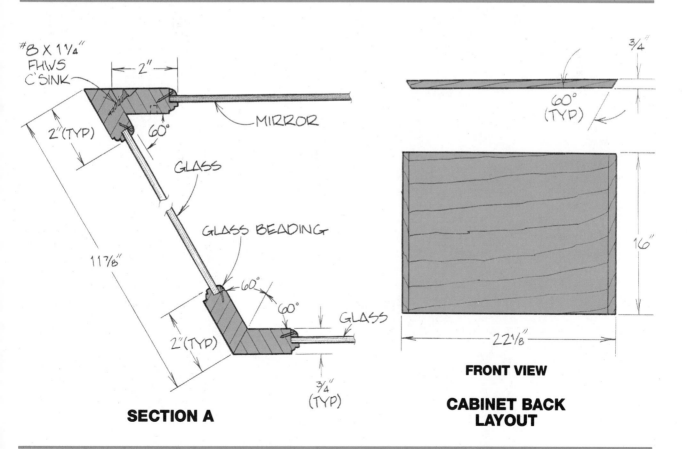

SECTION A

FRONT VIEW

CABINET BACK LAYOUT

8 Make the moldings and the glass beading.

Each of the moldings and the glass beading on the curio cabinet are made in a slightly different manner:

■ Fashion the *base moldings* with a shaper, molder, or table-mounted router. Make the shape in two separate passes, cutting the faces of the boards. First, cut the top corner with an ogee cutter. Complete the shape with a flute-and-quarter-round cutter. (See Figure 3.)

■ The *waist moldings* are also made in two passes, using the same two cutters — ogee and flute-and-quarter-round. (See Figure 4.) However, you must shape the edge of the stock instead of the face. Cut the shape in boards that are at least 3" wide, then rip the 1"-wide moldings free. *Never* try to shape slender stock; the wood may come apart in your hand.

■ Cut the *top moldings* on a table saw and a jointer. First, cut a cove in the face of the stock, using the table saw and an ordinary combination blade. Clamp a straightedge to the saw table, angled at 35° to the blade — this will serve as a guide fence. Adjust the depth of cut to ⅛". Pass the stock over the blade, keeping one edge firmly against the straightedge. Raise the blade another ⅛" and make another pass. Repeat this process until you have cut each cove ½" deep. (See Figures 5 and 6.)

Chamfer the edges of the stock on a jointer. Tilt the jointer fence at a 45° angle to the knives. Adjust the depth of cut to ⁵⁄₃₂". Joint all four corners, making *two* passes over the knives to cut the *front* corners, and *four* passes to cut the back corners. (See Figure 7.)

■ Cut the *glass beading* in a similar manner as the waist moldings. Shape the edge of ¼"-thick stock with a ¼" quarter-round cutter. Then rip the ⁹⁄₃₂"-wide beading free. Repeat until you have made all the beading you need.

Note: Make a little extra beading. These thin, narrow strips split easily.

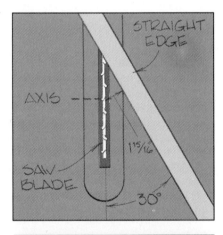

5/To cut coves in the top moldings, clamp a straightedge to the table saw. The face of the straightedge should be angled at 35° to the blade and positioned 1⁷⁄₁₆" from the axis, as shown.

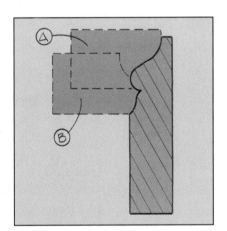

3/Make the base moldings in two passes, using an ogee cutter (A), then a flute-and-quarter-round cutter (B). Depending on the make of the cutters, you may not be able to make the exact shapes shown here. That's okay; the shape is not critical. Make something approximate.

6/Cut the cove in the stock, using the straightedge to guide the cut. Make several passes, cutting just ⅛" deeper with each pass. On the last pass, feed the wood very slowly to make the cut as smooth as possible.

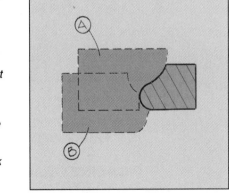

*4/Make the waist moldings in two passes with the same two cutters used to make the base moldings. However, cut the **edge** of the stock instead of the surface.*

*7/When you cut a chamfer with a jointer, always tilt the fence **toward** the knives. This prevents the wood from slipping when you cut.*

9 Fit and assemble the moldings.

Measure the cabinet before you cut the moldings to length. The dimensions of the case or the base may have changed slightly — this often happens with large projects. Make corresponding adjustments in the measurements of the moldings, and miter the ends at 60°.

Attach the top moldings to the case top with glue and screws, as shown in the *Top Molding Detail*. Fit the base and waist moldings to the cabinet, and mark the locations of the dowels. Drill ⅜"-diameter, ½"-deep holes in the moldings, and ⅜"-diameter, 1½"-deep holes in the cabinet. Attach both moldings with glue and dowels, and reinforce the base moldings with glue blocks.

Note: Some cabinetmakers prefer to cut moldings slightly long, then shave them to fit.

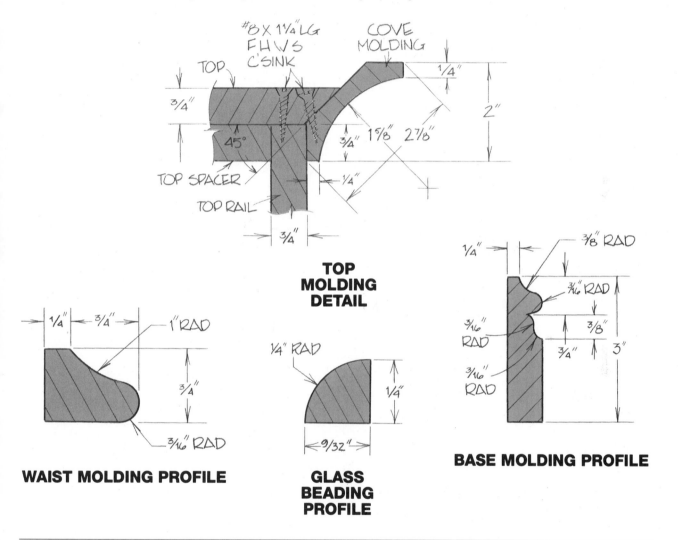

TOP MOLDING DETAIL

WAIST MOLDING PROFILE

GLASS BEADING PROFILE

BASE MOLDING PROFILE

10 Make and install the base panels.

Make the raised panels with a rotary planer, mounted in a drill press. Clamp a straightedge under the planer to guide the stock. Then cut 1⅛"-wide, ¼"-deep rabbets all around the *front* faces of the panels. To cut the corner radii, move the straightedge half the diameter of the planer *away* from the tool, and clamp a stop block to it. Position the block so it will halt the panel stock when the radii are perfect half circles. (See Figure 8.) Cut a radius in each corner of each panel.

8/When feeding stock into a rotary planer, follow the same safety rules that you do when using a table-mounted router: Feed the wood so the rotation of the planer helps hold it against the fence or straightedge.

Finish sand the completed panels and place them in the frames with the raised portions facing out. Secure them with flathead wood screws and flat washers, as shown in the *Panel Joinery Detail*.

Note: If you don't have a rotary planer and don't want to invest in one, make raised panels on a table

saw. Bevel the edges at 15°, using a tall fence extension to guide the stock. Adjust the position of the fence so the saw blade leaves a ⅛" step between the bevel and the flat area.

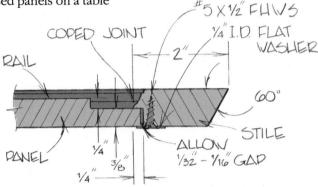

**PANEL
JOINERY
DETAIL**

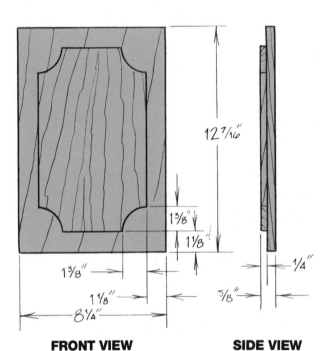

**BASE
PANEL
LAYOUT**

FRONT VIEW   **SIDE VIEW**

11 **Install the doors.** Mortise the side frames and the door frames for hinges, then hang the doors. Work carefully; the doors must hang precisely

on the hinges for the doors to close properly. When the doors are hung and working to your satisfaction, install pulls and magnetic catches.

TRY THIS! If you need to move a door slightly on its hinges, remove the screws from one-half of the hinge. (Which half doesn't matter.) Plug smaller screw holes with toothpicks and larger

holes with matchsticks. Redrill the pilot holes to one side or the other of the plugs, depending on which way you want the door to move. Then replace the screws.

12

Finish the curio cabinet. Detach the doors from the cabinet, and remove the panels and all hardware. Do any necessary touch-up sanding, then apply a finish to the completed cabinet. Be sure to apply as many coats to the inside of the cabinet as the outside. This will help prevent it from warping. Also finish the strips of glass beading.

After the finish dries, reassemble the panels and the doors. Cut glass beading to fit the frames, mitering the corners. Install a mirror in the case back frame and glass in the side frames and door. Hold the mirror and the glass in place with the beading, driving wire brads through the beading and into the frames as shown in *Section A*. Set the heads of the brads and touch up the finish on the beading.

Finally, put shelf supports in the holes in the case stiles. Lay the glass shelves on these supports. Most glass shops will make these trapezoid-shaped shelves on special order. Have them follow the *Shelf Layout*. The glass cutter should round the edges — especially the front edges — of the shelves so folks won't cut themselves when they reach inside the cabinet.

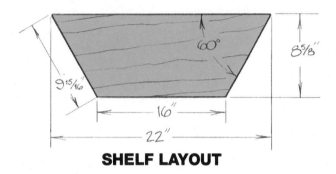

SHELF LAYOUT

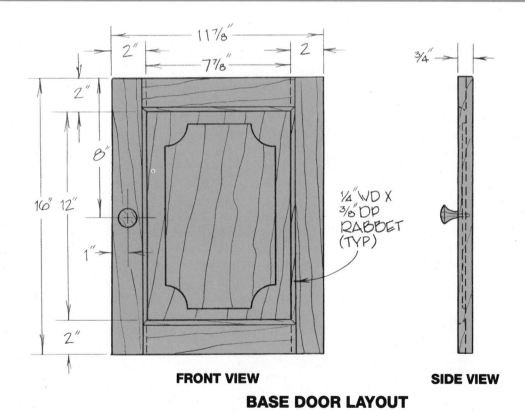

FRONT VIEW   **SIDE VIEW**

BASE DOOR LAYOUT

This project first appeared in the July/August 1982 issue of Hands On! magazine. It is reprinted by permission of Shopsmith, Inc.

STACKABLE/HANGABLE SHELVES

Stackable/Hangable Shelves

Sometimes simplicity is the mother of versatility. These simple display units have no glass doors or intricate frames. They're little more than elongated wooden boxes, but you can arrange them to store or display any small or medium-size item.

Furthermore, you can use these units almost anywhere. Hang them on a wall to make shelves, or stack them on a table, desk, or counter to create cubbyholes. Whether you stack them or hang them, they can be configured in dozens of different ways to blend with both contemporary and traditional settings.

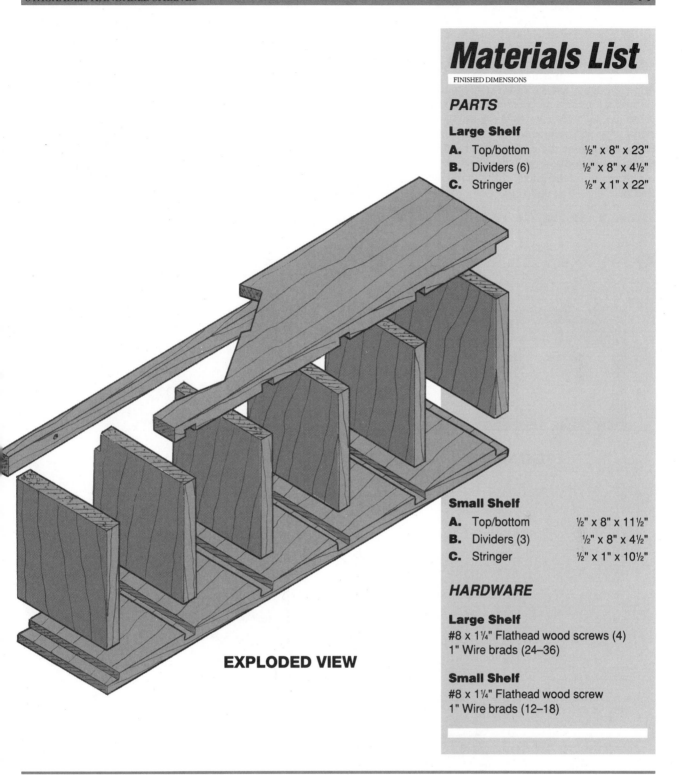

EXPLODED VIEW

Materials List

FINISHED DIMENSIONS

PARTS

Large Shelf

A. Top/bottom   ½" x 8" x 23"
B. Dividers (6)   ½" x 8" x 4½"
C. Stringer   ½" x 1" x 22"

Small Shelf

A. Top/bottom   ½" x 8" x 11½"
B. Dividers (3)   ½" x 8" x 4½"
C. Stringer   ½" x 1" x 10½"

HARDWARE

Large Shelf

#8 x 1¼" Flathead wood screws (4)
1" Wire brads (24–36)

Small Shelf

#8 x 1¼" Flathead wood screw
1" Wire brads (12–18)

---

1 ***Select and cut the stock.*** You'll need approximately 6 board feet of 4/4 (four-quarters) lumber to make the large shelf and 3 board feet to make the small one. The shelves shown were made from poplar, but you can use any cabinet-grade wood for the tops, bottoms, and dividers. Use hardwood for the stringers, since these parts must support the weight of the shelves and their contents.

When you have chosen the wood, plane it to ½" thick. Cut the parts to the sizes in the Materials List.

2 *Cut the joinery in the top and bottom.*

Lay the top and bottom on your workbench, edge to edge, with the ends flush. Measure the position of the dadoes and rabbets on one part, then mark the joints across both parts. (See Figure 1.) By marking both parts at once, you ensure that all the joints match perfectly and that all the dividers will be straight up and down when you assemble the shelf.

After marking the joints, cut ½"-wide, ¼"-deep rabbets and dadoes with a dado cutter or a router.

1/Use a framing square to mark across both the top and the bottom.

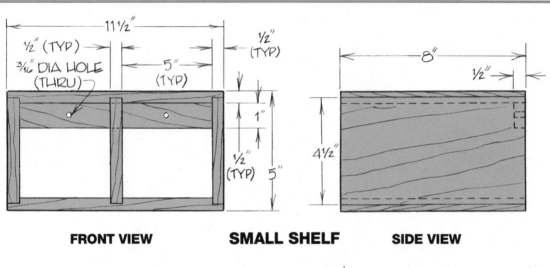

FRONT VIEW   **SMALL SHELF**   **SIDE VIEW**

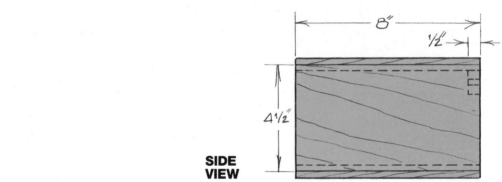

SIDE VIEW

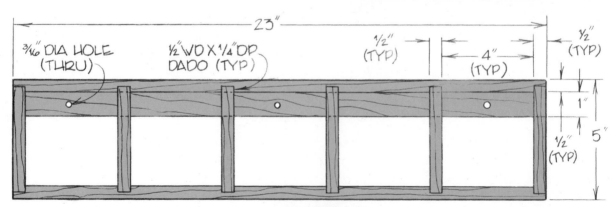

FRONT VIEW
LARGE SHELF

3 Notch the dividers. Using a band saw or a saber saw, cut 1¼"-wide, ½"-deep notches in the back top corners of the *middle* dividers, as shown in the *Middle Divider Layout*. Do not notch the left- or right-most dividers.

TRY THIS! To save time, stack the middle dividers and tape them together. Mark the notch on the top divider only, then cut the entire stack.

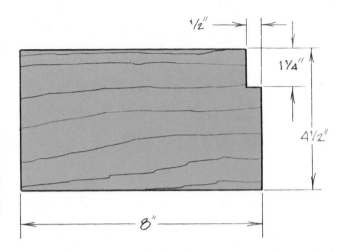

MIDDLE DIVIDER LAYOUT

4 Drill mounting holes in the stringer. Drill two or three ¼"-diameter holes through the stringer at regular intervals. Position these holes where you can reach them from the front of the assembled shelf — they mustn't be blocked by a divider. You'll use these holes to mount the shelf to a wall.

5 Assemble the shelf. Finish sand all the parts, being careful not to round over any adjoining surfaces. Assemble the top, bottom, and dividers with glue, then reinforce the glue joints with wire brads. Set the heads of the brads below the wood surface. Remember, you must orient all the notches in the middle dividers at the upper back corner of the assembly.

Glue the stringer to the top and the middle dividers. The back face of the stringer must be flush with the back edges of the top and the dividers. Reinforce the glue joints with flathead screws, driving them through the stringer and into the middle dividers. Countersink the heads of the screws so they're flush with the wood surface.

6 Finish the assembled shelf. When the glue dries, sand all joints clean and flush. Do touch-up sanding on any other surfaces that need it. Finish the shelf, applying an equal number of coats to both the inside and the outside. Either use a finish that penetrates the surface, like tung oil or oil stain, or paint the shelf. Avoid finishes that you have to rub out, such as varnish. It would be very difficult to smooth the finish inside the small compartments.

7 Hang the shelf (optional). If you want to hang the shelf, position it on a wall. Insert an awl through each hole in the stringer, marking the location of the holes on the wall. Remove the shelf and find the studs in the wall. If any of the hole marks are over studs, drill ³⁄₁₆" pilot holes into the studs at those points.

Install Molly anchors at the locations of the other holes.

Put the shelf back in place on the wall. Drive #12 roundhead wood screws through the holes over the studs and ³⁄₁₆" bolts into the Molly anchors.

Note: To hang this project on a masonry wall, use expandable lead anchors and lag screws.

Glass-Top Display Stand

This contemporary display stand is a unique blend of old and new craftsmanship. Its three-legged design has a long history. Originally, furniture makers built pieces like this to sit solidly on uneven floors. However, they soon found the design had other merits: It was light, strong, and elegant. So, although floors have leveled out over the centuries, three-legged pieces have proliferated — wine tables, sewing stands, occasional tables, plant displays. The woodworkers of each era have experimented with three-legged designs for every kind of small table.

The glass top itself doesn't have as much history behind it. Glass-top tables have only been common since the invention of tempered glass, early in this century. However, glass imparts a distinctive lightness and elegance that have quickly made it popular among contemporary craftsmen. It was inevitable that a glass top would be coupled with three legs.

This marriage of old and new traditions was built by Mark Burhans of Athens, Ohio. Mark is the proprietor of Burhans Woodworks and sells his wood-turnings throughout the country. He's also a part-time administrator at the prestigious Dairy Barn Gallery, where he helps put together the "American Contemporary Works in Wood" show every two years. ❋

Materials List

FINISHED DIMENSIONS

PARTS

A.	Legs (3)	1¼" dia. x 36⅛"
B.	Rungs (9)	⅝" dia. x 11¼"
C.	Feet (6)	1¾" dia. x 1⅛"
D.	Finials (3)	2" dia. x 1½"

**EXPLODED
VIEW**

HARDWARE

11½" dia. x ¼" Tempered glass (2)

1

Select and cut the stock. To make this table, you need just one board foot of 4/4 (four-quarters) stock for the rungs, one board foot of 8/4 stock for the feet, one board foot of 10/4 stock for the finials, and three 2" x 2" x 36" turning squares for the legs. You can use any cabinet-grade wood as long as it's straight and free of loose knots, checks, and other defects, and as long as you use a hard wood for the feet. Mark used a variety of woods — cherry for the rungs, hard maple for the feet, spalted maple for the finials, and oak for the legs.

Rip the 4/4 stock into 1" x 1" x 12" turning squares, the 8/4 stock into 1⅛" x 2" x 2" squares, and the 10/4 stock into 2½" x 2½" x 2¼" squares. When you make the squares for the feet and the finials, cut and sand the *ends* flat. The sides can remain rough cut, but the ends must be smooth. Cut several extra squares of each size in case you make a mistake while turning.

Note: The wood grain of each turning square must be *parallel* to the turning axis. It's very difficult to turn wood when the grain is perpendicular — the chisels dig into the stock.

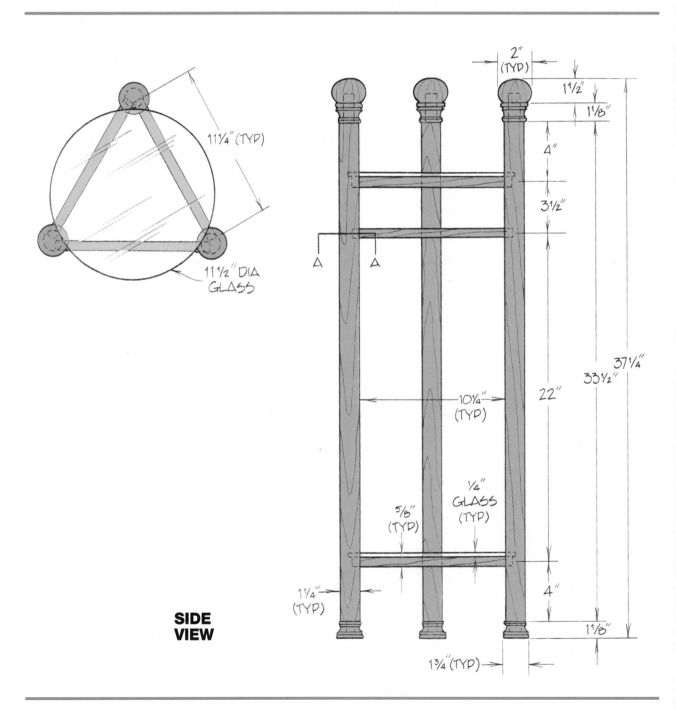

**SIDE
VIEW**

2 Turn the legs and rungs.

Turn the legs and rungs. Both the legs and the rungs are straight turnings with no beads or coves. However, you must turn them to precise diameters. The rungs must be ⅝" in diameter, the legs 1¼", and the tenons on the ends of the legs ¾". To help turn these diameters, make calipers from a scrap of plywood or hardwood as shown in the *Calipers Layout*. Use the calipers to gauge the turnings as you make them. (See Figures 1 and 2.)

1/Turn the stock, checking the diameter with the calipers. Stop when the calipers slip over the wood, up to the ¹⁄₁₆" step. This indicates that you are approaching the final diameter.

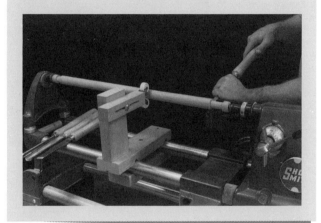

TRY THIS! If the legs wobble or whip on the lathe, use a steadyrest to stabilize them.

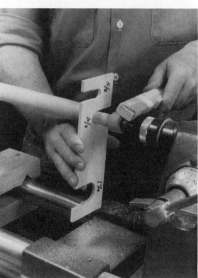

2/With the calipers in one hand and 80# sandpaper in the other, sand the last bit of stock from the turning. (Sandpaper removes stock more slowly than a chisel. This helps prevent you from removing too much stock.) Stop when the calipers slip past the step.

Finish sand the turnings on the lathe, but be careful not to remove too much more stock. Don't sand the tenons or the ends of the rungs at all, or they may not fit properly in the mortises. Cover them with masking tape to prevent accidentally removing stock as you sand.

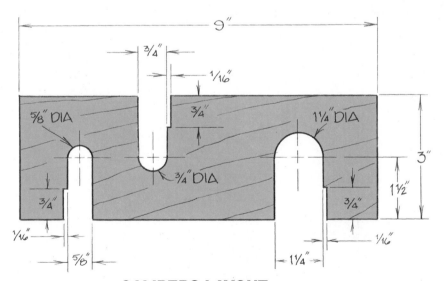

CALIPERS LAYOUT

3 Turn the feet and the finials.

The feet and the finials are pressure-turned between two blocks of wood. Cut these blocks from scrap hardwood, as shown in the *Pressure Turning Blocks* drawing. Drill a ¾"-diameter, ½"-deep hole, centered in the face of each block. Then drill a ⅛"-diameter hole through the center of the larger holes. Also cut two ¾"-diameter dowels — one 1⅛" long (to turn the bottom feet and the finials), and the other 2" long (to turn the top feet).

Drill a ¾"-diameter, ¾"-deep hole in one end of each turning block for the bottom feet and the finials. Drill ¾"-diameter mortises completely through the top feet blocks.

To mount a turning block on the lathe, first insert the proper length of ¾"-diameter dowel in the mortise. Then put a pressure block over each end of the turning block, fitting the dowel into the stopped hole in one (or both) of the pressure blocks. Mount this assembly between the lathe centers, inserting the points of the centers into the ⅛"-diameter holes in the pressure blocks. Advance the tailstock to keep all three blocks pressed together tightly while you turn the one in the middle.

Turn the feet and finials to the shapes shown in the *Foot Detail* and *Side View*. If you wish, cut templates from scraps of ¼" plywood or hardboard. Use these to help make all the feet and finials the same. (See Figure 3.) Finish sand each part on the lathe.

3/To help duplicate the feet and the finials, make templates from scraps of plywood or hardboard. As you turn, compare the turning to the template. Try to match the shapes as closely as possible.

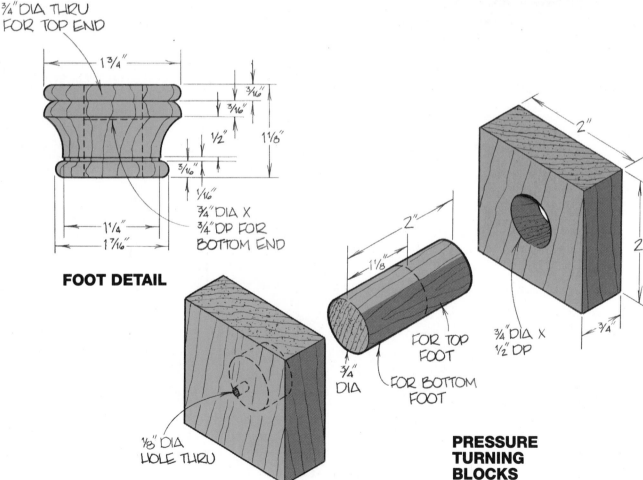

¾" DIA THRU FOR TOP END

FOOT DETAIL

¾" DIA X ¾" DP FOR BOTTOM END

⅛" DIA HOLE THRU

FOR TOP FOOT

¾" DIA

FOR BOTTOM FOOT

¾" DIA X ½" DP

PRESSURE TURNING BLOCKS

4 **Mortise the legs.** You must drill the round mortises in the legs at an *offset* angle from the center of the stock, as shown in *Section A*. This is difficult to do without a special guide — the drill bit wanders. To prevent this, make the *Mortising Jig* shown. Glue up a block from hardwood scraps, and bore a 1¼"-diameter hole through the length. Drill a ⅝"-diameter, 2"-deep hole down from the top face to intersect the larger hole. Scribe a short line on one end of the block, next to the larger hole, to use as an indexing mark. This line must be parallel to the centerline of the ⅝"-diameter hole.

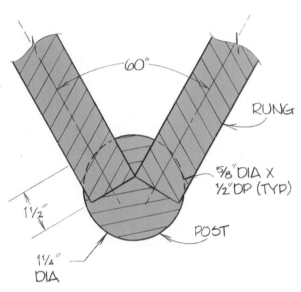

SECTION A

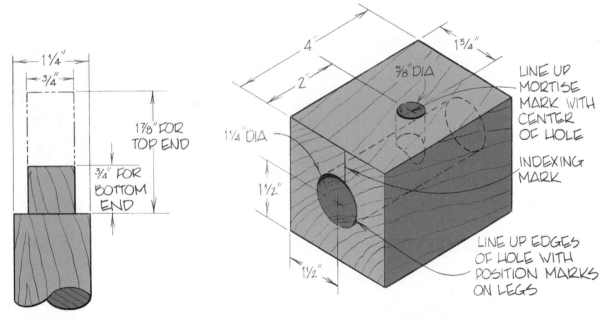

**TENON
DETAIL**

MORTISING JIG

Also make a *Mortising Gauge* from a 2½" x 2½" square of heavy paper or posterboard. Draw a 1¼"-diameter circle centered on the paper, then lay out the mortise centerlines as shown. Cut the circle out of the paper, leaving two lines a short distance apart on the circumference of the hole.

You must make six round mortises in each leg, two rows of them with three mortises in each. To locate these rows, insert each leg through the mortising gauge and mark both mortise centerlines. (See Figure 4.) Then place the leg in a long, V-shaped groove — a doorjamb works well — so you can use one edge of the groove as a straightedge. At each of the centerline marks, scribe a line down the length of the leg. (See Figure 5.)

Measure along the leg to locate the mortises in each row. Put two more marks beside each of the mortise marks, one 2" above it and the other 2" below, as shown in the *Leg Layout*. These are positioning marks — you'll use them to position the leg in the mortising jig.

Mount a ⅝"-diameter brad-point bit in a drill press. (This bit must be *very* sharp.) Position the mortising jig directly under the bit with the ⅝"-diameter hole facing up. Adjust the drill stop to halt the bit when it's extended 2" into the jig.

Insert a leg through the 1¼"-diameter hole and turn it so both sets of mortise marks are facing *up*. Sighting

through the ⅝"-diameter hole, slide the leg through the jig until you can see a mortise mark. Align one of the mortise rows with the indexing mark on the jig. (The other mortise row must still face up. It should be to one side of the indexing mark, not below it.) Also align the positioning marks even with the ends of the jig.

Drill the mortise, taking care the leg doesn't slide or turn in the jig. (See Figure 6.) Repeat this procedure for all the mortises in the row. Then turn the leg end for end and drill the mortises in the second row. Repeat for each leg.

5/Use the "V" created by the moldings in a door-jamb to mark the mortise rows down the length of each leg. To keep from getting pencil lead on the doorjamb, stick masking tape over it.

4/Make a paper gauge to locate the rows of mortises on each leg. Scribe just **one** set of marks on each leg.

6/To keep a leg from moving in the jig when you drill a round mortise, clamp it in a V-jig.

5 **Assemble the stand.** Dry assemble the stand to test the fit of the parts. When you're satisfied that everything goes together properly, take it apart and reassemble it with glue. Clamp the assembly by fastening band clamps around all three legs. Wipe away any excess glue with a wet rag.

Cut an 11½"-diameter disk out of a scrap of ¼" plywood and place it on the stand where you will

put either shelf. There should be just a tiny (1/32") gap between the edge of the disk and each leg. If the disk is too large, trim off a little stock. If it's too small, make a new disk.

When you've properly sized the disk, have a glass shop use it as a template to make two round shelves of ¼"-thick tempered glass. Also have them sand the edges of the shelves to remove any sharp corners.

6 **Finish the stand.** Do any necessary touch-up sanding on the stand and apply a finish. Mark chose to paint several of the wooden parts on the piece and leave others natural. You can paint the parts, stain them, or finish them according to your tastes. After the finish dries, lay the glass shelves in place.

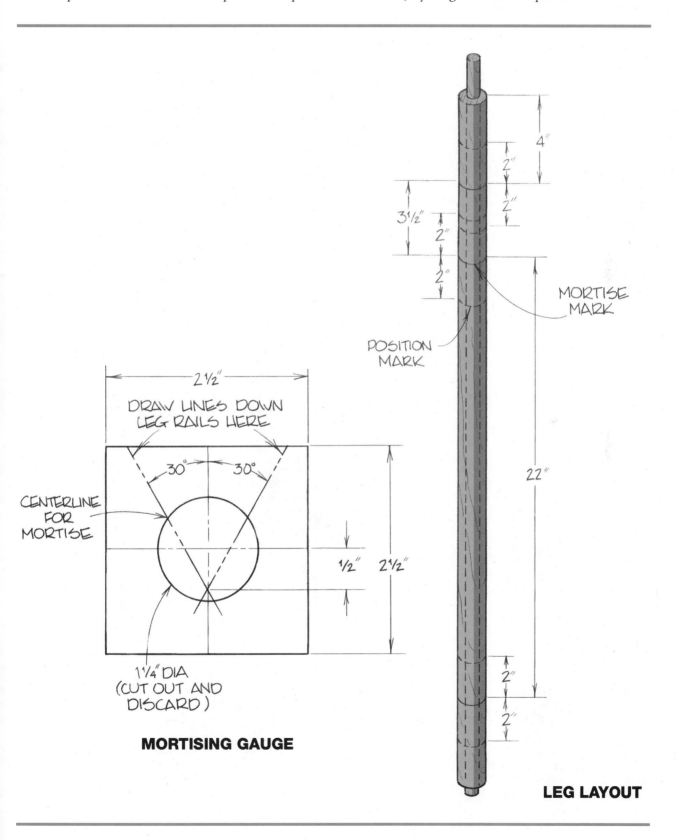

MORTISING GAUGE

LEG LAYOUT

Corner Showcase

orners are great
places to display collec-
tibles, since they're often
empty. The reason that
they're empty, however,
is there aren't many dis-
play cases — or any other
furniture pieces — that fit
them. If you place a rec-
tangular case in a corner,
the front always faces to
one side or the other,
making it difficult to see
the contents.

This hanging showcase
is an exception. Even
though it's rectangular, it
has no sides — just two
backs and two fronts. The
backs are solid wood; the
fronts are glass doors. The
wooden backs mount on
the walls, and the doors
face out to the room so
you can easily view the
contents. Both doors
open, providing access
from either side, and the
shelves rest on movable
pins, so they can be
raised or lowered.

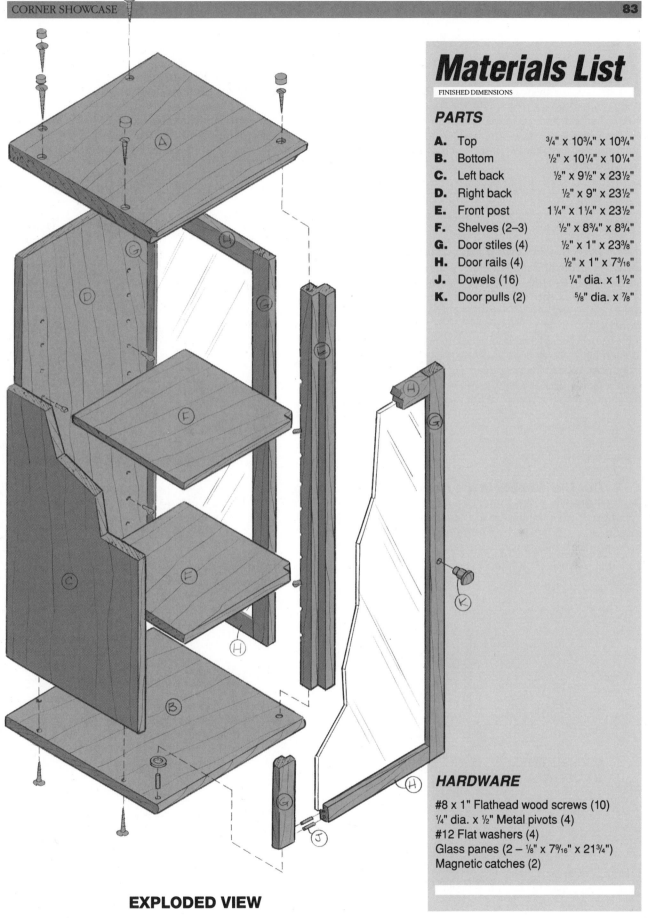

EXPLODED VIEW

Materials List

FINISHED DIMENSIONS

PARTS

A.	Top	¾" x 10¾" x 10¾"
B.	Bottom	½" x 10¼" x 10¼"
C.	Left back	½" x 9½" x 23½"
D.	Right back	½" x 9" x 23½"
E.	Front post	1¼" x 1¼" x 23½"
F.	Shelves (2–3)	½" x 8¾" x 8¾"
G.	Door stiles (4)	½" x 1" x 23⅜"
H.	Door rails (4)	½" x 1" x 7³⁄₁₆"
J.	Dowels (16)	¼" dia. x 1½"
K.	Door pulls (2)	⅝" dia. x ⅞"

HARDWARE

#8 x 1" Flathead wood screws (10)
¼" dia. x ½" Metal pivots (4)
#12 Flat washers (4)
Glass panes (2 – ⅛" x 7⁹⁄₁₆" x 21¾")
Magnetic catches (2)

1 **Select and cut the stock.** You can build this project from almost any cabinet-grade wood. For a contemporary look, make the case from a light-colored wood, such as maple, birch, or pine. (The case shown is made of white pine.) For a more traditional appearance, use a darker wood such as cherry, walnut, or mahogany.

You'll need about 8 board feet of 4/4 (four-quarters) lumber and a 2 x 2 turning square at least 24" long. Plane one board foot of the 4/4 stock to ¾" thick and the remainder to ½". Plane the 2 x 2 to 1¼" square. Cut all the parts to the sizes in the Materials List, except the door pulls. You'll turn these later from scraps of ¾" stock.

2 **Drill the shelving support holes.** Measure and mark the locations of the shelving support holes on the back pieces, as shown in *Section A.* Using one of the back pieces as a measuring stick, mark corresponding holes on the *inside corner* of the front post.

Use a drill press to make a ¼"-diameter, ⅜"-deep hole at each mark. To drill the shelving support holes in the front post, mount the post in a V-jig with the inside corner facing up. (See Figure 1.) Drill the post holes diagonally into the corner of the post.

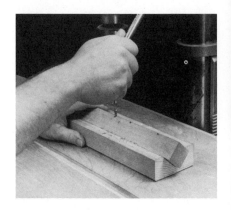

1/Hold the post in a V-jig to drill holes in the corner. To help control splintering, tape the corner and mark the holes directly on the tape. Peel the tape off after you've made the holes.

3 **Cut the rabbets in the front post.** Using a table-mounted router or a dado cutter, make ½"-wide, ½"-deep rabbets in the *outside faces* of the front post, as shown in the *Front Post Detail.* The rabbets should be in opposite corners.

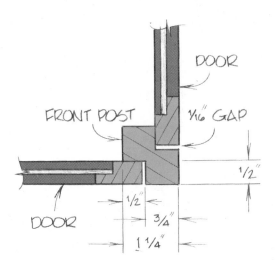

CROSS SECTION

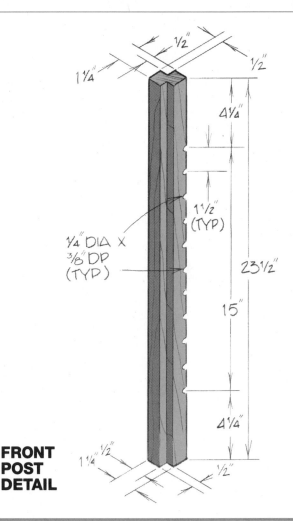

FRONT POST DETAIL

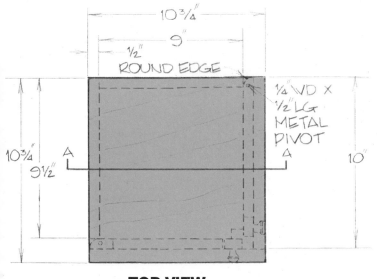

10¾"

9"

½"
ROUND EDGE

¼" WD X ½" LG METAL PIVOT

A

A

10¾"

9½"

10"

TOP VIEW

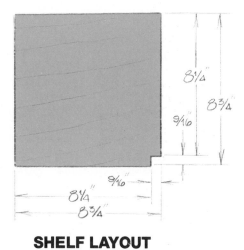

8¼"

8¾"

9/16"

8¼"

8¾"

9/16"

SHELF LAYOUT

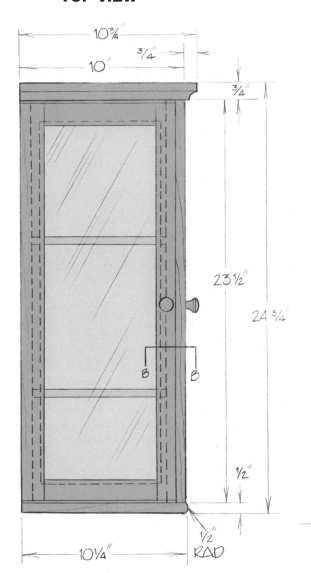

10¾"

10"

¾"

¾"

23½"

24¾"

B

B

½"

½" RAD

10¼"

END VIEW

SECTION A

6¼"

1¼"

4¼"

1½" (TYP)

23½"

¼" DIA X 3/8" DP (TYP)

½" (TYP)

4¼"

½"

9"

4

Shape the top edge. Cut a cove in the *outside edge and end* of the top, using a router and a ½"-radius cove bit. Sand the cove to remove any mill marks.

> *TRY THIS!* When you shape end grain, make the shape in multiple passes to keep the bit from "hogging" or tearing the wood. Lower the bit just ⅛" – ¼" with each pass. Also feed the router across the stock as quickly as you can, without stopping or slowing. This will help keep the wood grain from burning.

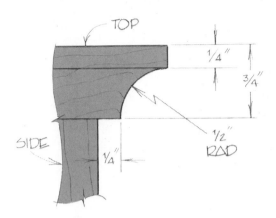

TOP EDGE DETAIL

5

Assemble the door frames. Lay out the door parts and mark the locations of the dowels. Using a doweling jig, drill a ¼"-diameter, ¾"-deep dowel hole at each mark. Also drill ¼"-diameter, ¼"-deep holes in the two *front* door stiles for the door pulls. Assemble the rails and stiles with dowels and glue.

Let the glue dry for at least 24 hours. Sand the dowel joints clean and flush, then clamp the door frames to your workbench. Using a piloted rabbeting bit, rout ¼"-wide, ¼"-deep rabbets around the *inside edge* of each door. (These rabbets will hold the glass panes.) Square the corners of the rabbets with a hand chisel.

6

Drill the pivot holes. Mark the locations of the pivot holes on the top, bottom, and door frames. Drill ¼"-diameter, ⅜"-deep holes at each mark. Take care that each pivot hole is precisely perpendicular to the wood surface. Use a drill press to make

the holes in the bottom and top, and a doweling jig when drilling the door frames. Round the *inside pivot corner* of each door with a file or a small hand plane, as shown in the *Pivot Detail*.

7

Assemble the case. Finish sand all parts and assemblies — top, bottom, backs, posts, and door frames. Then assemble the parts in the following order:

■ Attach the back pieces with glue and flathead wood screws, driving the screws through the left back and into the edge of the right back. Countersink the screws.

■ Attach the top to the back pieces with glue and screws, driving the screws through the top. Counterbore *and* countersink the screws, then cover the heads with wooden plugs.

■ Place metal pivots and washers in the holes in the door frames. Mount the doors in the top, then put the

bottom in place. Use screws (but *not* glue) to attach the bottom to the back pieces. Countersink the screws, but don't cover the heads — you'll need to remove the bottom later. Check the action of the doors, making sure they turn freely on their pivots.

■ Place the front post between the top and bottom. Experiment with different positions until you find one that looks straight. The doors should fit flush in the rabbets with just a tiny gap, as shown in *Section B*. Fasten the top of the post with glue and a screw. Counterbore *and* countersink this screw, then cover the head with a wooden plug. Fasten the bottom of the post with a screw. Just countersink the screw; don't glue the bottom of the post or cover the screw head.

8 **Turn and install the door pulls.** Cut a ¾"
x ¾" x 3½" turning block from some of the ¾"-
thick scrap. Turn the door pulls on a lathe, following
the *Door Pull Detail*. If you don't have a lathe, use a
drill press to make the turnings. (See Figure 2.) Glue the
pulls in the front door frame stiles.

*2/You can make
small turnings on
a drill press with
this simple jig. To
mount the stock
in the drill chuck,
drive a #12 screw
part way into one
end. Cut the head
off the screw, then
clamp the shank in
the chuck.*

9 **Finish the showcase.** Unscrew the bottom
from the case, and remove the door frames and
pivots. Replace the bottom while you finish the completed project; this will hold the post steady.

Do any necessary touch-up sanding, then apply a
finish to all wooden surfaces. Be sure to apply as many
coats to the back surfaces and the inside of the case as
you do to the parts that will show. This will help prevent the case from warping. When the finish dries, rub
it smooth and apply a coat of paste wax.

Install glass in the door frames, holding it in place
with glazier points. Remove the bottom, then reassemble the doors, pivots, and bottom. Finally, mount
magnetic catches on the post and front door stiles.

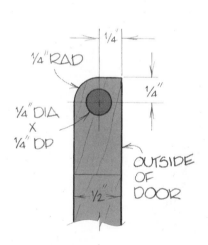

PIVOT DETAIL

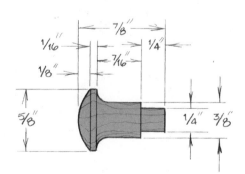

DOOR PULL DETAIL

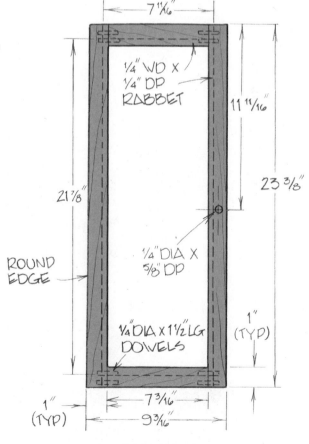

DOOR LAYOUT

Multiple Picture Frame

When you process a roll of film, you sometimes find one exceptional photo that's too good to hide in a scrapbook. You want to display it where you can enjoy it often. Roll after roll, these special photos accumulate and, before long, you have a whole collection of them.

This "multiple picture" frame provides an attractive way to organize and display your special snapshots. The frame is just a large board with many small cutouts. Each cutout can be a different size and shape; just choose the cutout that best frames a particular photo. To put a photo in the frame, simply loosen six screws and remove the back. ❋

Materials List

FINISHED DIMENSIONS

PARTS

A. Frame   3/4" x 16" x 20"

B. Top/bottom
spacers (2)   1/8" x 1/2" x 18"

C. Side spacers (2)   1/8" x 1/2" x 15"

D. Back   1/4" x 15" x 19"

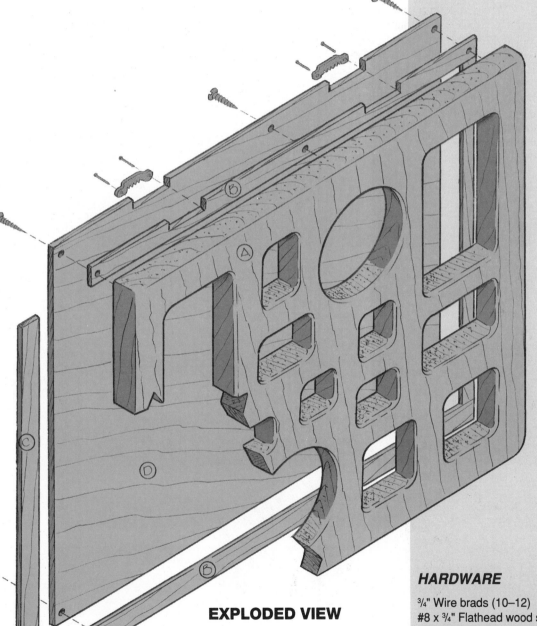

EXPLODED VIEW

HARDWARE

3/4" Wire brads (10–12)
#8 x 3/4" Flathead wood screws (6)
1/8" x 137/8" x 177/8" Glass pane
Serrated picture hangers (2)

1 Select and cut the stock.

Select and cut the stock. To make this project, you need approximately 3 board feet of 4/4 (four-quarters) stock and a scrap of ¼" plywood at least 15" x 19". The frame shown was made out of oak, but you can make it out of any cabinet-grade wood. Choose straight lumber with *no* warps, cups, or twists. If the wood is distorted at all, the frame won't lie flat against the wall.

When you have chosen the wood, let it stand in your shop for several weeks to become acclimated to the temperature and humidity. Then plane it to ¾" thick and glue up the stock you need to make the frame. Cut the frame to size. Plane the scraps to ½" thick, and rip ⅛"-wide strips from them to make the spacers. Also cut the plywood back to size.

TRY THIS! When making a project that you don't want to cup or warp, build it from *quarter-sawn* or *rift-sawn* lumber. Because of the manner in which this wood is cut from the tree, it will not distort as easily as plain-sawn lumber. You can tell quarter-sawn and rift-sawn lumber by looking at the end grain — the annual rings are nearly perpendicular to the face of the board. In plain-sawn lumber, the rings curve parallel to the face.

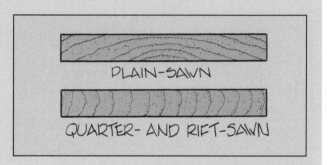

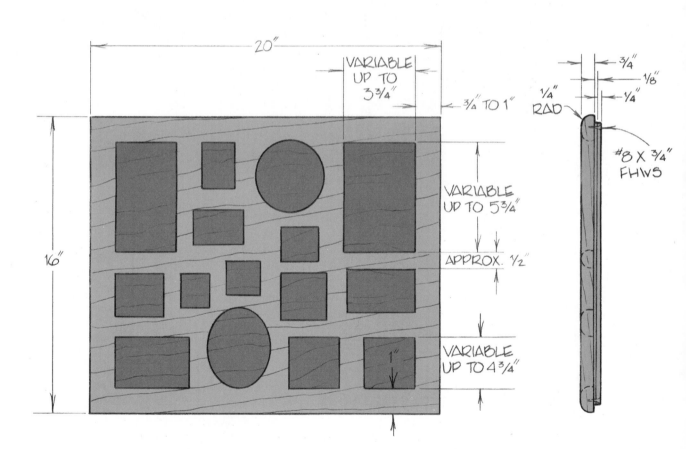

FRONT VIEW   **SIDE VIEW**

2 Lay out the cutouts on the frame.

Consider the sizes and shapes that you want to make the cutouts. Most snapshots are 4" x 6", or smaller, so the largest cutouts should be 3¾" x 5¾". You can make the cutouts rectangular, square, round, or oval. You'll probably want a variety of shapes and sizes.

Once you've decided on the cutouts you want, lay them out on the frame. There are two ways to do this: You can place them at *random*, or you can use a *grid*.

The easiest method is to arrange them at random. (The multiple picture frame shown was laid out in this manner.) Cut the sizes and shapes of the cutouts you want out of stiff paper. Place these paper templates on the face of the frame, arranging them and rearranging them until you come up with a pleasing layout. Try not to bunch up all the same sizes or shapes in the same area. Also, try to keep the spacing between the templates as uniform as possible. When you've made a

good arrangement, tape them down to the wood. Trace around each template with a pencil.

The grid method is more time-consuming, but it produces a more orderly and often more aesthetic layout. First, draw a grid with gutters and squares like the one shown in the *Sample Grid* drawing. The overall grid should be slightly smaller than the frame. The squares should be the *minimum* size that you want to make a cutout, and the gutter should be the space you want to maintain between cutouts. (In the sample, the squares are 1⅝" to a side, and the gutters are ½" wide.)

Lay out the cutout shapes on the grid, making sure that *all four* sides of each cutout touch the sides of squares. Always use the gutters to separate the cutouts, and never split a square or a gutter. (The *Alternate Layout* was done on a grid.) When you finish the layout, use carbon paper to transfer it to the frame stock.

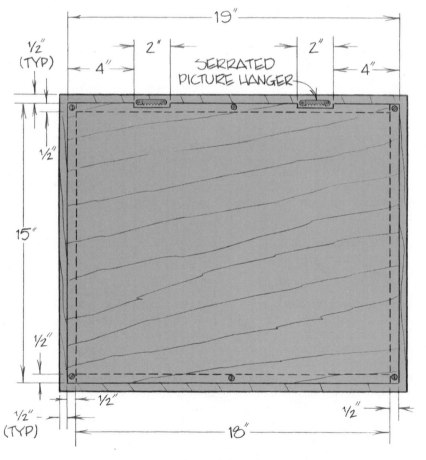

BACK VIEW

3 **Saw the cutouts.** Wherever you've marked the wood for a cutout, drill a ½"-diameter hole inside the lines. Insert the saber saw or a scroll saw through the hole, saw to the lines, then cut out the waste. Sand or file the sawed edges to remove the saw marks.

TRY THIS! Saw the cutouts with fine-cut saw blades that have as many teeth per inch as you can find. These cut more slowly than ordinary blades, and they break more easily, but they leave a smoother edge. This will make the project easier to sand.

4 **Round over the frame.** Clamp the frame securely to your workbench. Using a hand-held router and a ¼"-radius quarter-round bit, round over the front edges of the frame, as shown in the *Side View*. Rout around the inside edge of each cutout, as well as the outside edge of the overall frame. Be very careful when routing those areas where the frame is only ½"

wide. If you press too hard against the wood, the frame may split.

Carefully sand the front surface of the frame, removing all mill marks and burn marks left by the router. Lightly sand the back surface to remove any splinters or chips around the edges of the cutouts.

5 **Attach the spacers to the frame.** Lay out the spacers on the back of the frame, parallel to the edges and ends and ½" in. Mark their positions. Using a band saw or saber saw, notch the top spacer to fit around the picture hangers. Glue the top and bottom spacers in position, and attach the side spacers with wire brads.

Note: The grain direction of the side spacers opposes that of the frame. If you glue them in place, the glue will restrict the frame from expanding and contracting with changes in temperature and humidity. This will cause the frame to warp or split. The brads, however, don't restrict movement. They bend slightly when the frame shrinks or swells.

6 **Finish the frame.** Do any necessary touch-up sanding on the frame, then apply a finish. Be sure to apply as many coats to the front surface as you

do to the back. This, too, will help prevent the frame from warping.

7 **Mount the photos and assemble the frame.** Put the back in place and trace the shapes of the cutouts. Remove the back, then arrange photographs on it. Place each photo over the outline of a cutout, making sure it completely covers the outline. If the photo overlaps another outline, trim it. The photos must not overlap each other; however, they should cover all the outlines. (You shouldn't be able to see any pencil lines when you're finished with the arrangement.) When the photos are all positioned properly, tape them to the back.

Place a pane of glass inside the spacers. Then lay the back in place on top of the spacers. Holding the back securely to the frame, pick up the assembly and look at it from the front. Inspect the photos to make sure they're all positioned properly. When you're satisfied that they are, attach the back to the spacers with flathead wood screws. Install serrated picture hangers near the top of the frame, in the notches, as shown in the *Back View*. Hang the frame from two small nails or screws, driven partway into the wall.

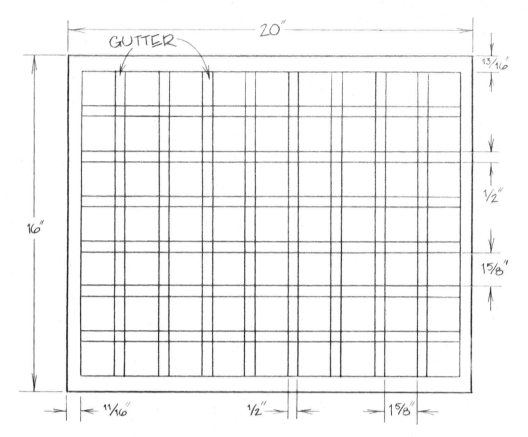

SAMPLE GRID

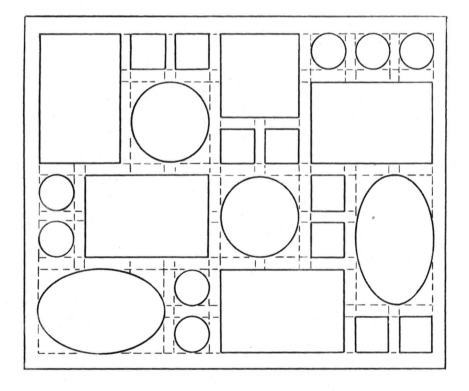

ALTERNATE LAYOUT

Corner Cupboard

Our ancestors often displayed their china and other breakables in corner cupboards. Corners are good locations for keeping fragile items, since they are outside a room's normal traffic pattern. Whatever you put in a corner won't be in the way, and it's less likely to be knocked over or broken. You can enjoy it without bumping into it every time you pass through the room.

The corner cupboard shown is triangular to make the best use of the corner space. Despite its odd shape, it's simple to build. The case is assembled with ordinary butt and miter joints, and the shelves rest in dadoes. This particular cupboard is open to the room, so you can easily see and reach the contents from almost any angle. The simple fretwork front and appliqué make an attractive frame in which to display china — and many other collectibles.

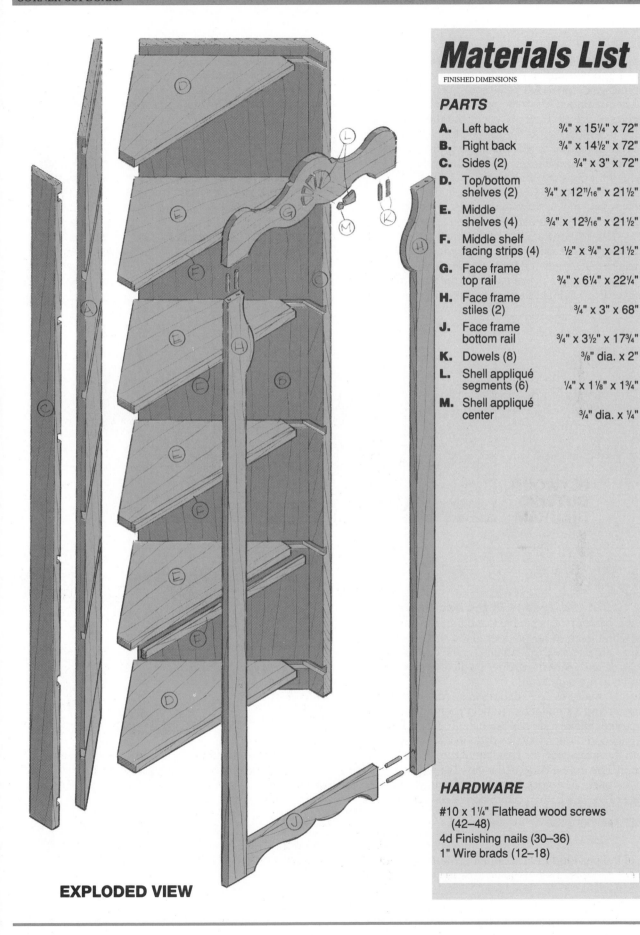

Materials List

FINISHED DIMENSIONS

PARTS

A.	Left back	3/4" x 151/4" x 72"
B.	Right back	3/4" x 141/2" x 72"
C.	Sides (2)	3/4" x 3" x 72"
D.	Top/bottom shelves (2)	3/4" x 1211/16" x 211/2"
E.	Middle shelves (4)	3/4" x 123/16" x 211/2"
F.	Middle shelf facing strips (4)	1/2" x 3/4" x 211/2"
G.	Face frame top rail	3/4" x 61/4" x 221/4"
H.	Face frame stiles (2)	3/4" x 3" x 68"
J.	Face frame bottom rail	3/4" x 31/2" x 173/4"
K.	Dowels (8)	3/8" dia. x 2"
L.	Shell appliqué segments (6)	1/4" x 11/8" x 13/4"
M.	Shell appliqué center	3/4" dia. x 1/4"

HARDWARE

#10 x 11/4" Flathead wood screws (42–48)

4d Finishing nails (30–36)

1" Wire brads (12–18)

EXPLODED VIEW

1 Select the stock and cut the parts.

Select the stock and cut the parts. The cupboard shown is made from a single sheet of cabinet-grade plywood and 8 board feet of 4/4 (four-quarters) lumber. The wide parts — back pieces and shelves — are cut from plywood, and the remainder is made from solid stock. If you want to make this project entirely from solid wood, purchase approximately 32 board feet of lumber, and glue up the wide stock needed for the shelves and backs.

Plane the wood to ¾" thick, then cut the backs, shelves, and facing strips to the sizes in the Materials List. (You'll cut the other parts as you assemble the case.) Miter the front edges of the back pieces and the rear edges of the sides at 45°, as shown in the *Top View*. If you're using plywood for the wide parts, follow the *Plywood Cutting Diagram* to make the back pieces and shelves from a single sheet. If you're using solid wood

exclusively, cut all six shelves to the measurements given for the top and bottom shelves (part D). Don't bother making the facing strips (part F), since you won't need to face the middle shelves.

> **TRY THIS!** When cutting large sheets of plywood, lay a 4' x 8' sheet of fiberboard (sometimes called builder's board) on the floor of your shop, then place the plywood over it. Adjust the depth of cut on a circular saw so the blade cuts through the plywood and ¹⁄₁₆" to ⅛" into the fiberboard. This method is safer and easier than the traditional practice of laying the plywood across sawhorses, because it provides more support.

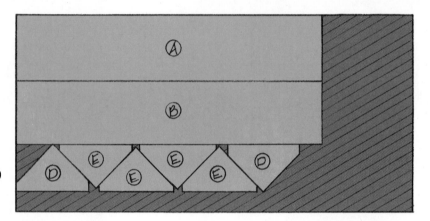

PLYWOOD CUTTING DIAGRAM

2 Cut the dadoes in the backs and sides.

Cut the dadoes in the backs and sides. Lay the back and the side parts across the workbench or on the floor of the shop. Turn the inside surfaces face up and line up the bottom edges. Measure and mark the dadoes on all four pieces, as shown in the *Side View*.

Cut ¾"-wide, ⅜"-deep dadoes in all four parts, using a router and a straight bit. To help position and guide the router, make a simple T-square from two lengths of plywood. Rout a practice dado in a scrap, cutting both the scrap *and* the crossbar of the T-square. Then, simply line up the dado in the crossbar with the marks on the backs and sides. Butt the crossbar against the edge of the stock, then clamp both the stock and the

1/When routing dadoes or rabbets, use a homemade T-square to position and guide the router. Make this jig from cabinet-grade plywood so it will remain straight and square.

T-square to the workbench. Rout the dado, keeping the base of the router pressed against the T-square arm. (See Figure 1.) Make each dado in several passes, cutting ⅛" to ¼" deeper each time.

> **TRY THIS!** To keep the veneered surface of the plywood from splintering when you rout it, put a single layer of masking tape over the areas you want to cut. Mark the joints on this tape, then remove it after you've made the cuts.

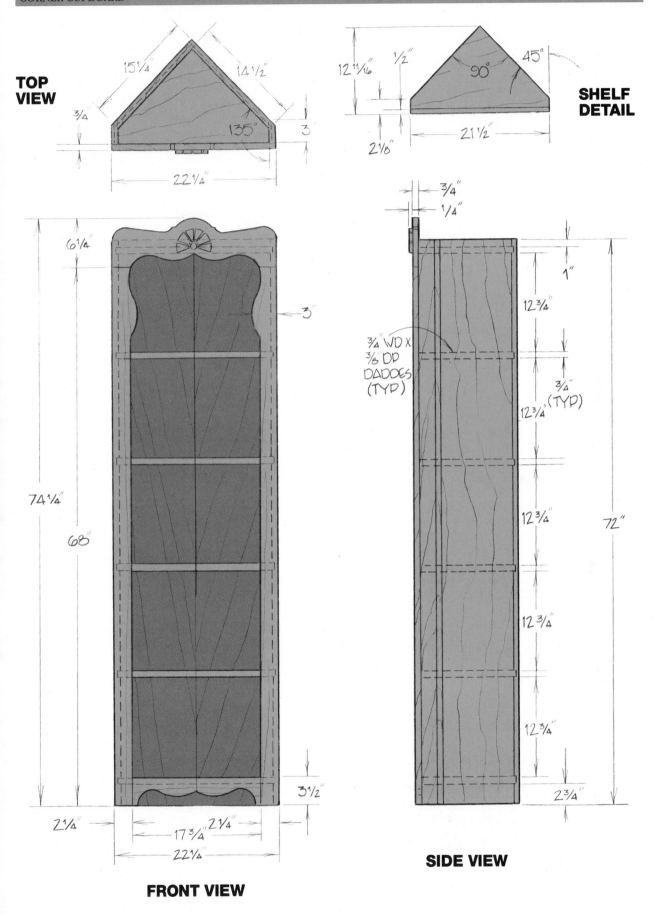

TOP VIEW

SHELF DETAIL

FRONT VIEW

SIDE VIEW

3 Face the middle shelves.

Face the middle shelves. If you've cut the wide parts from plywood, then you must face the middle shelves with strips of solid wood. Otherwise, the plies will show on the assembled cupboard.

Glue the facing strips to the front edges of the shelves and tack them in place with wire brads. Set the heads of the brads.

4 Assemble the case.

Assemble the case. Finish sand the backs, sides, and shelves. If you've made any of these parts from plywood, be careful not to sand through the veneered surface. Dry assemble the parts to test their fit. When you're satisfied that they fit properly, assemble them with glue and flathead wood screws in this order:

■ Fasten the left back to the right back, driving screws through the left back and into the right. Make sure the dadoes line up precisely.

■ Slide the shelves into the dadoes. The shelves should be seated against the bottoms of the right *and*

left dadoes. Fasten the shelves in place, driving screws into them from the backs. Tighten the screws, drawing the edges of the shelves tight against the bottom of the dadoes.

■ Fasten the sides in place with finishing nails. Drive the nails through the sides and into the shelves or backs.

Note: Countersink all screws so the heads are flush with the wood surface. Set the heads of the nails *below* the surface, and cover them with putty or stick shellac.

Let the glue dry on the assembled case, then sand all joints clean and flush.

5 Make the face frame.

Make the face frame. Measure the front of the case. Chances are the dimensions will have changed slightly from those in the drawings. This is normal for large pieces; it's difficult for even the most experienced woodworkers to make long, wide boards behave properly. Adjust the measurements of the face frame parts to correspond with the assembled case, then cut the parts to size.

Temporarily join the rails and stiles with dowels. (If you have a plate joiner, you can also use wooden plates.) Don't glue the parts together yet.

Enlarge the *Face Frame Pattern* and trace it onto the rails and stiles. Disassemble the parts, and cut the shapes with a band saw or saber saw. Sand the sawed edges.

FACE FRAME PATTERN

1 SQUARE = 1"

6 Attach the face frame to the case.

Attach the face frame to the case. Finish sand the rails and stiles, then assemble them with dowels and glue. Before the glue dries, attach the frame to the front of the case with glue and finishing nails. Set the heads of the nails below the surface of the wood and cover them. After the glue dries, sand all joints clean and flush.

TRY THIS! To give the corner cabinet a rustic country look, use square-cut nails instead of finishing nails to attach the face frame. Drill pilot holes for the nails so they won't split the stock. Set the heads flush with the surface and don't cover them.

7

Make and attach the appliqué. The top rail is decorated with a traditional shell or "rising sun" appliqué. Plane a piece of scrap stock to ¼" thick to make the parts for this appliqué, then cut them with a scroll saw or jigsaw. You can cut all six segments at once, following the procedure shown in Figures 2 through 5.

Lay the cupboard down on the floor and brace it so the face frame is level. Carefully position the parts of the appliqué on the top rail, as shown in the *Shell Appliqué Layout*. Make small, light pencil marks to indicate the positions of the parts, then glue them to the top rail.

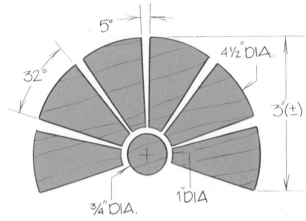

SHELL APPLIQUÉ LAYOUT

2/To make the segments, first cut six pieces ¼" x 2" x 3". Place the pieces on top of each other and tape the stack together. Mark the shape of a segment on the top piece, then cut just one side.

3/Sand the cut side of the stack on a belt sander or disk sander to remove the saw marks.

4/After you've sanded away the saw marks, retape the side to secure the stack.

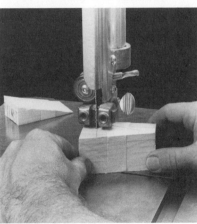

*5/Cut the next side, sand it, and retape it. Repeat for all four sides. At no time should you leave more than one side of the stack untaped. Following this procedure, you'll make all six segments **precisely** the same.*

8

Finish the cupboard. Allow the glue on the appliqué to dry for at least 24 hours. Then finish sand the appliqué, slightly rounding the edges and corners. Carefully remove any visible pencil marks from the surface of the top rail with fine sandpaper.

Do any necessary touch-up sanding to the other parts of the cupboard, then apply a finish. The cupboard shown is painted with traditional milk paint. The outside is a dark barn red, while the appliqué and the inside are a lighter, brighter shade of the same red. Cupboards such as these were traditionally painted a light color on the inside to make it easier to see the objects on the shelves.

Round and Oval Frames

Not every picture looks best in a rectangular frame. Some appear better when set inside an oval or round picture frame. These frames add a homey, intimate ambience to the subject. They can also make a picture look old and treasured.

You can make both round and oval frames from a single piece of wood, shaped with a router or shaper. They can be almost any size, although the frames shown are designed to hold standard-size photos or artwork.

Materials List

FINISHED DIMENSIONS

PARTS

Oval Frames

A. For 8" x 10" photograph   ¾" x 10½" x 12½"

B. For 5" x 7" photograph   ¾" x 7" x 9"

C. For 4" x 5" photograph   ¾" x 5¾" x 6¾"

Round Frames

A. For 8"-dia. artwork   ¾" x 10½" dia.

B. For 5"-dia. artwork   ¾" x 7" dia.

C. For 4"-dia. artwork   ¾" x 5¾" dia.

HARDWARE

⅛" Glass or clear acrylic plastic pane (to fit each frame)

½" Wire brads or glazing points (4–6 per frame)

Posterboard or cardboard (to fit each frame)

Small eye screws (2 per frame)

10-lb. Picture-hanging wire (8"–12" per frame)

---

1 Select and cut the stock. Since each of these frames is made from a single piece of wood, select stock that doesn't split easily. The frames shown are poplar, but you can also use cherry, walnut, maple, oak, and birch. Avoid soft or brittle woods, such as pine or cedar. Also avoid woods with knots, checks, cracks, or other defects. Many craftsmen prefer to use quarter-sawn or rift-sawn lumber for projects like these, since this stock does not warp as easily as plain sawn lumber.

Once you choose the wood, plane it to ¾" thick and glue up the widths needed. Cut the stock a little oversize — about ¼" longer and wider than needed — to allow room for error.

---

2 Lay out the frame on the stock. To lay out a *round* frame, simply draw two concentric circles on the stock with a compass. The inside circle marks the frame opening, and the outside marks the circumference. The difference in diameters should be 1½" or more, so the frame will be at least ¾" wide. Narrower frames are too fragile. The round frames shown in the drawings are all 1" wide or wider.

To lay out an *oval* frame, you must mark two concentric ovals. To do this, you need a compass, a straightedge, and a string. Mark the outside oval first, then the inside, following the procedure shown in Figures 1 through 5.

As with the round frames, the differences in the length and width of the ovals should be 1½" or more,

making the frames no less than ¾" wide. The oval frames shown in the drawings are all 1" wide or wider.

As long as the frame isn't too narrow, you can make the ovals any size you want. The oval frames in the drawings are designed for three standard-size photos — 8" x 10", 5" x 7", and 4" x 5". The *Oval Frame Sizes* chart shows the axes measurements for each frame opening (inside oval) and the overall dimensions (outside oval). Another chart — *Oval Layout Information* — shows the distance between the foci for all the ovals in the first chart. It also tells the circumference of the string (after tying the ends together) needed to draw the ovals.

Oval Frame Sizes

Photo Size	Frame Opening	Overall Dimensions
8" x 10"	7½" x 9½"	10½" x 12½"
5" x 7"	4½" x 6½"	7" x 9"
4" x 5"	3¾" x 4¾"	5¾" x 6¾"

Oval Layout Information

Oval Sizes	Foci Distance	String Circumference
10½" x 12½"	7⅝"	20⅝"
7½" x 9½"	6¹³⁄₁₆"	15⁵⁄₁₆"
7" x 9"	5⅝"	14⅝"
4½" x 6½"	4¹¹⁄₁₆"	11⅛"
5¾" x 6¾"	3½"	10¼"
3¾" x 4¾"	2¹¹⁄₁₆"	7¹³⁄₁₆"

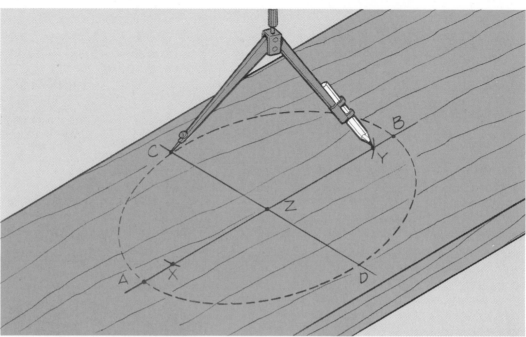

1/Mark the length or **major axis** (AB) and the width or **minor axis** (CD) of the outside oval on the stock. The axes must be perpendicular to each other and cross at the center (Z). Adjust a compass to half the length of the major axis (AZ or ZB). Put the point of the compass on one end of the minor axis (C or D), and scribe arcs that intersect the major axis (at X and Y). These are called **foci.**

2/Drive two tacks or small nails at the foci (X and Y). Put a third tack at one end of the major axis (A or B) so all three tacks are in a line.

3/To determine the length of the string, loop it around the line of tacks, and tie the ends together in a square knot.

4/Remove the tack at the end of the major axis (A or B) but leave the string and the other tacks in place. Put the point of a pencil through the square knot and stretch the string taut. Draw the oval, keeping the string taut and pulling it around the **tacks.**

5/Superimpose the major and minor axes of the inside oval on the axis lines of the outside oval. Then repeat the procedure to draw the inside oval.

3 Cut the opening in the frame.

Drill a hole through the waste, inside the smaller circle or oval. Insert a saber saw or scroll saw blade through this opening and saw to the inside marks. Cut out the waste.

> **TRY THIS!** If you keep the entry hole small and drill it near the line you want to cut, you can use the waste to make a smaller frame. For example, you can cut the 4"-diameter artwork frame from the waste of the 8"-diameter frame. Likewise, the 4" x 5" oval photo frame can be made from the waste of the 8" x 10" frame.

Sand the inside edge of the circle or oval to remove the saw marks. A drum sander, mounted in a drill press, helps speed this operation. (See Figure 6.) Be careful to maintain a "fair curve" as you sand. A fair curve is a cabinetmaker's term that describes a curve with no indentations or flat spots — just a smooth, flowing line. To preserve this curve, keep the piece moving as you sand it. Don't let the sander dwell on any one spot too long.

Note: Do *not* cut the outside curve of the frame yet. The extra stock provides a good, safe handhold while you cut the rabbet and shape the frame during the next few steps. It will also help hold the frame together while you mill it.

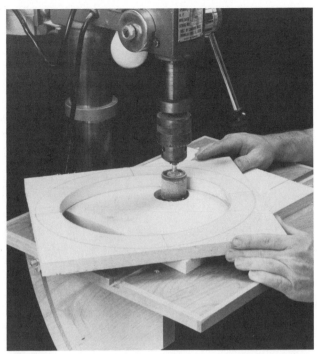

6/To use a drill press and drum sander to sand the inside of a cutout, first drill a hole in a piece of scrap. This hole should be slightly larger than the drum, and the scrap should be larger than the piece you want to sand. Fasten the scrap to the drill press table, with the hole directly under the drum. Place the piece on top of the scrap, with the cutout over the hole. Lower the drum through the cutout and partway through the hole, then lock it in place. Turn the drill press on and sand the piece.

4 Cut a rabbet in the back of the frame.

Using a piloted rabbeting bit, cut a ¼"-wide, ¼"-deep rabbet in the back inside edge of the frame. (See Figure 7.) Because you are cutting end grain during part of this operation, make the rabbet in several shallow passes. Cut just ¹⁄₁₆" – ⅛" deeper with each pass. Be careful to feed the stock (or to move the router) *against* the rotation of the bit.

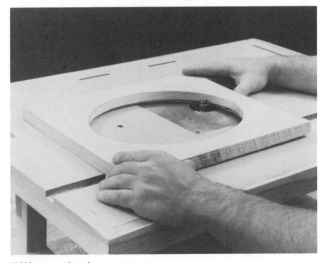

7/When routing the frame, you can use either a hand-held or table-mounted router. Here, a table-mounted router is used to cut the rabbet.

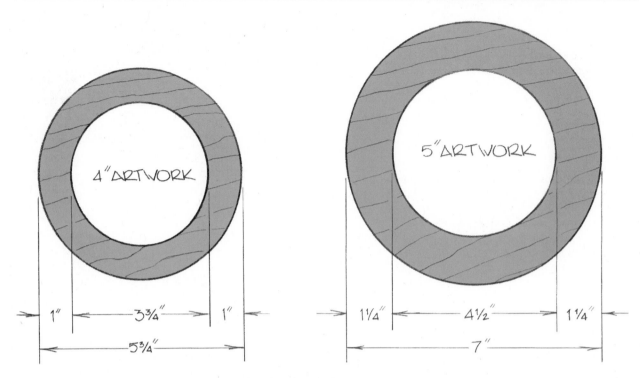

ROUND FRAMES

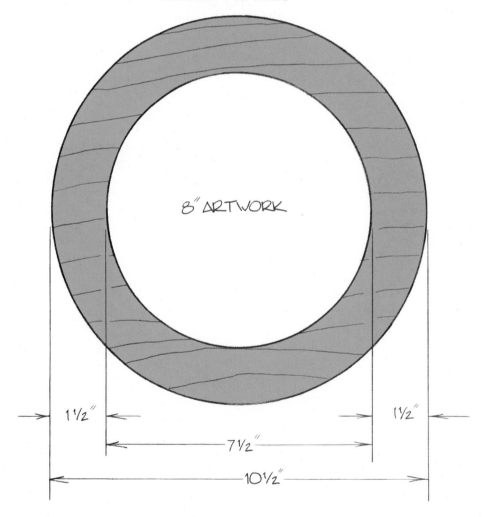

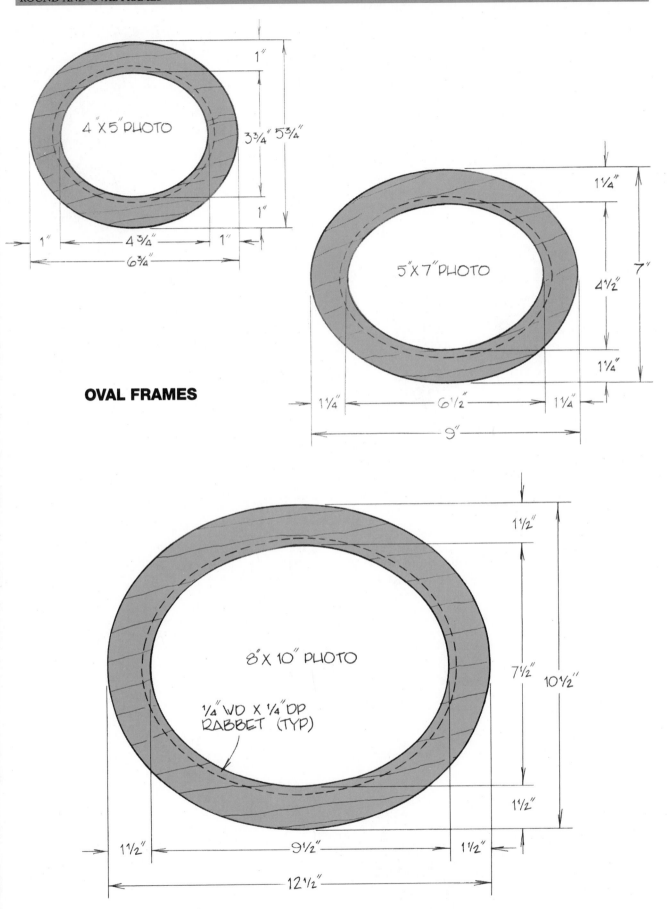

OVAL FRAMES

4″ X 5″ PHOTO

1″

3¾″ 5¾″

1″

1″ 4¾″ 1″

6¾″

5″X7″PHOTO

1¼″

7″

4½″

1¼″

1¼″ 6½″ 1¼″

9″

8″X 10″ PHOTO

¼″ WD X ¼″ DP RABBET (TYP)

1½″

7½″ 10½″

1½″

1½″ 9½″ 1½″

12½″

5

Shape the front of the frame. Change the rabbeting bit for a piloted shaping bit — ogee, bead, cove, round-over, chamfer, or some other shape you fancy. Several possibilities are shown in the *Frame Profiles*. Then turn the frame over and rout the shape in the front inside edge. (See Figure 8.)

Note: You must make the half-round shape in the *Frame Profiles* in two steps. First, round the inside edge. Then, after you cut the outside curve, round the outside edge. Perform the last step very carefully, particularly if you are shaping a small frame. If you rout too fast or press too hard, the frame may come apart in your hands. To help prevent this, make the shape in several passes, routing just $\frac{1}{16}$" deeper with each pass.

8/If you use a hand-held router for rabbeting or shaping, be sure the frame stock is secured to the workbench.

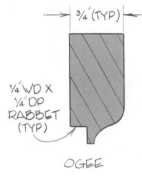    

OGEE   BEAD   COVE   HALF-ROUND   CHAMFER

FRAME PROFILES

6

Cut the outside curve. Using a scroll saw, saber saw, or band saw, cut the outside curve of the frame. Sand the sawed edge with a drum sander or disk sander to remove the saw marks. Once again, be careful to preserve a fair curve.

TRY THIS!   To make it easier to achieve a fair curve, cut a little wide of the cut line, then sand up to it.

7

Sand and finish the frame. Finish sand the frame, being careful not to soften or blend any of the hard edges you want to preserve on the shaped portion. Then apply a finish. The frames shown were painted with many different colors of paint. The colors were overlaid with brushes, sponges, or fingers, producing the effects shown. You can also use stains or clear finishes, if you wish.

8

Install the glass and the hardware. Have round or oval panes of glass made at a glass shop to fit the frames. Or cut your own panes from clear acrylic plastic. Cut the picture and a posterboard or cardboard backing to fit the frame. Carefully clean the pane to remove any dirt; then install the pane, picture, and backing in the frame. Secure them with glazing points or small brads.

Drive two small eye screws into the back side of the frame, one on either side, slightly above center. Stretch picture-hanging wire between these screws, then hang the frame from a picture hook.

Tool Display Cabinet

Many craftsmen collect antique tools. And what better place to display a collection than in a tool cabinet? This particular cabinet opens up to create a large exhibit with three different types of storage — shelves, racks, and drawers. When closed, it's an elegant display of craftsmanship.

You can also use the cabinet in your shop, of course, where the display serves a different purpose. With all of your important tools neatly arranged in a single cabinet, it's easy to find a tool when you need it. The cabinet holds the tools at a comfortable height, so you can reach them without bending or stretching. And if you keep this cabinet near the work area, it will save you steps. You'll spend less time fetching the tools and more time working with them. ●

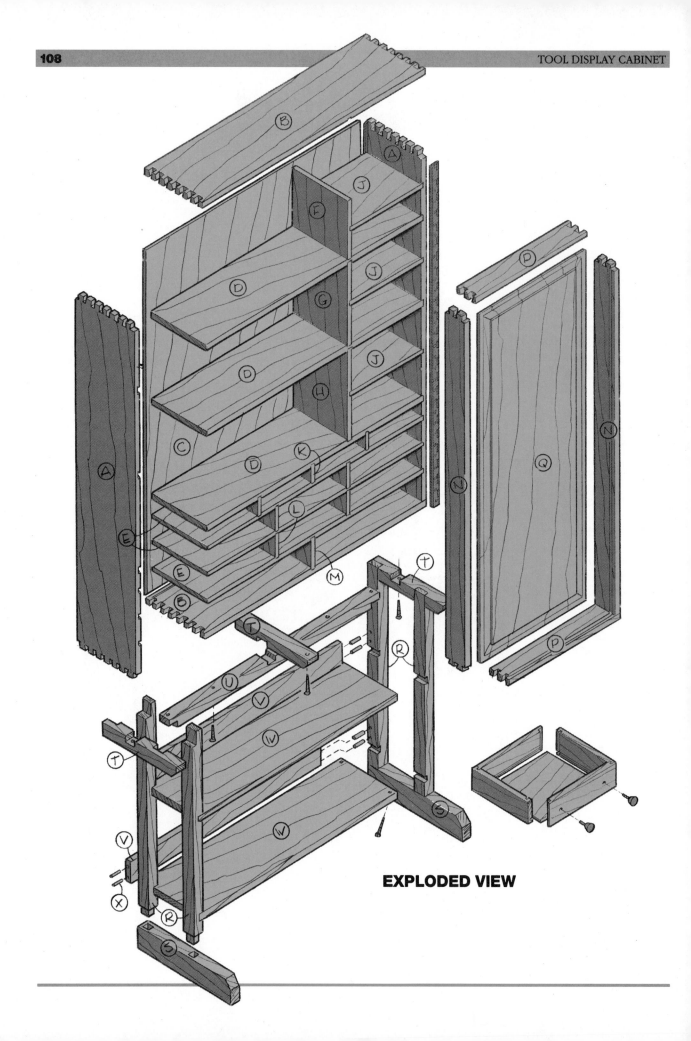

EXPLODED VIEW

Materials List

FINISHED DIMENSIONS

PARTS

A.	Case sides (2)	¾" x 9" x 48"
B.	Case top/bottom (2)	¾" x 9" x 36"
C.	Case back (plywood)	¼" x 35¼" x 47¼"
D.	Long shelves (3)	¾" x 8½" x 35"
E.	Drawer partitions (3)	½" x 8½" x 35"
F.	Upper shelving divider	½" x 8½" x 8½"
G.	Middle shelving divider	½" x 8½" x 11½"
H.	Lower shelving divider	½" x 8½" x 12½"
J.	Short shelves (3)	½" x 8½" x 11¹¹⁄₁₆"
K.	Upper drawer dividers (3)	½" x 8½" x 2⅜"
L.	Middle drawer dividers (4)	½" x 8½" x 3⅜"
M.	Lower drawer divider	½" x 8½" x 4⅛"
N.	Door sides (4)	¾" x 3" x 48"
P.	Door tops/bottoms (4)	¾" x 3" x 18"
Q.	Door panels (2)	¾" x 17" x 47¼"
R.	Legs (4)	1½" x 1½" x 28"
S.	Feet (2)	1½" x 3" x 15"
T.	Supports (3)	1½" x 1½" x 12"

U.	Stretcher	1½" x 1½" x 34½"
V.	Backstops (2)	¾" x 3" x 31½"
W.	Bottom shelves (2)	¾" x 9" x 32½"
X.	Dowels (8)	⅜" dia. x 2"
Y.	Small drawer fronts/backs (8)	½" x 1¹⁵⁄₁₆" x 8⅛"
Z.	Small drawer sides (8)	½" x 1¹⁵⁄₁₆" x 8"
AA.	Small drawer bottoms (4)	¼" x 7¾" x 7⅝"
BB.	Medium side drawer fronts/backs (8)	½" x 2¹⁵⁄₁₆" x 11⅛"
CC.	Medium side drawer sides (8)	½" x 2¹⁵⁄₁₆" x 8"
DD.	Medium side drawer bottoms (4)	¼" x 7¾" x 10⅝"
EE.	Medium middle drawer fronts/backs (4)	½" x 2¹⁵⁄₁₆" x 10⅞"
FF.	Medium middle drawer sides (4)	½" x 2¹⁵⁄₁₆" x 8"
GG.	Medium middle drawer bottoms (2)	¼" x 7¾" x 10⅜"
HH.	Large drawer fronts/backs (4)	½" x 3¹¹⁄₁₆" x 16⅞"
JJ.	Large drawer sides (4)	½" x 3¹¹⁄₁₆" x 8"
KK.	Large drawer bottoms (2)	¼" x 7¾" x 16⅜"

HARDWARE

1½" x 46½" Piano hinges and mounting screws (1 pair)
#12 x 2" Flathead wood screws (6)
#10 x 1¼" Flathead wood screws (4)
¾" Wire brads (10–12)
Hook-and-staple catches and mounting screws (4)
Small drawer pulls (24)

Starting Up

1

Modify the design to suit your needs.
Whether it holds antique, "retired" tools, or tools that are still very much in use, a tool cabinet is a very personal item. Every woodworker collects a unique set of tools according to his or her needs and interests. The cabinet must hold *your* tools, organized in a way that *you* find practical and aesthetic. Because of this, you may need to change the size of the cabinet, the arrangement of the shelves and drawers, or some other part of the design. Here are some important points to consider:

Size — As designed, the cabinet is not intended to hold every tool in your shop, nor will it hold an extensive tool collection. It will display a medium-size collection of hand tools or, if you use it in your shop, the important hand-held tools that you use most often. To determine the size of cabinet you need, gather all the tools you want to display and see how much room they take up; also decide how much more room you'll need to expand your collection. Adjust the size of the cabinet accordingly.

Depth — You can change the width and the height without interfering with the function of the cabinet, but you may be asking for trouble if you increase the depth. Once you make a cabinet more than one or two tools deep, you create an organizational nightmare. In a deep cabinet, you can't see the tools on the back row without moving tools in front of them. Over time, the cabinet becomes more and more disorganized. Keep the cabinet shallow so you can easily see all the tools.

Reach — If you will be using the cabinet in your shop, raise or lower the stand as necessary so you won't have to bend or stretch to reach any tool. Both the highest shelf and the lowest drawer should be within an arm's length. Even if you're using the cabinet just for display, the tools should be easily accessible. You'll want to take them down to dust them, maintain them, or just admire them from time to time.

Fixed or adjustable storage — You can make the configuration of shelves in a cabinet permanent or temporary. The shelves shown here are permanent. However, if you wish to rearrange the inside of the cabinet from time to time, mount the shelves on adjustable supports or standards.

Drawers — Some tools are best stored in drawers. In this cabinet, one drawer holds a set of knives for the rabbet plane; another holds drill bits for the brace. Still others hold tools that, because of their small size or odd shape, can't be hung up or placed on a shelf. Decide which tools you want to store in drawers, then calculate the number and size of drawers needed.

When planning the drawers, keep them as shallow as possible. The same logic applies here as did to the cabinet: If a drawer is more than one or two tools deep, you'll find yourself shuffling tools to reach the one you want. The contents of a deep drawer quickly become a jumble. If you must make deep drawers, build shallow stacking trays to fit inside them. These will help keep the contents organized.

2 **Select and plane the stock.** Most tool cabinets are made from hard, durable woods. Maple and birch are the traditional choices, but you can also use cherry, walnut, oak, and ash. The cabinet shown is mostly birch. The door panels and drawer fronts are curly red oak, and the cabinet back and drawer bottoms are birch-veneered plywood. Avoid soft woods like pine and mahogany. These quickly become worn and dented as you take tools out of the cabinet and return them.

To make the cabinet as designed, you'll need approximately 70 board feet of 4/4 (four-quarters) lumber, 7 board feet of 8/4 lumber, and a 4' x 8' sheet of cabinet-grade ¼" plywood. Plane 45 board feet of the 4/4 stock to ¾" thick, and the remaining 25 board feet to ½" thick. Plane all the 8/4 stock to 1½" thick.

Note: If you use figured woods, as shown, check that the planer knives are very sharp, and remove just ¹⁄₃₂" of stock at a pass.

Making the Case

3 **Cut the case parts.** Although this is not a complex project to build, there are a lot of parts. To help simplify construction, divide the project into five subassemblies — case, stand, shelves, drawers, and doors. Work on one subassembly at a time, beginning with the case.

Also remember that lots of parts can mean lots of opportunities for inaccuracy. Each error will make it progressively more difficult to fit the parts and the subassemblies together. To keep your cuts accurate, check the alignment and the adjustment of all your tools *before* you begin this project. When you're satisfied they are set up properly, cut the parts of the case. Check the tool settings periodically as you work.

TRY THIS! One of the most common causes of inaccurate cuts is *tool deflection* — the blade bends or wobbles slightly as you use it. To help prevent this, make all of your important cuts in two steps. Rip or crosscut a board to within ¹⁄₁₆" (about half a blade width) of its final dimension, then cut away the last bit of stock with a second cut. The blade will deflect less on the second pass because it's only cutting a tiny amount of wood. To further control deflection, feed the stock slowly and evenly. The "whir" of the blade shouldn't change pitch during the cut.

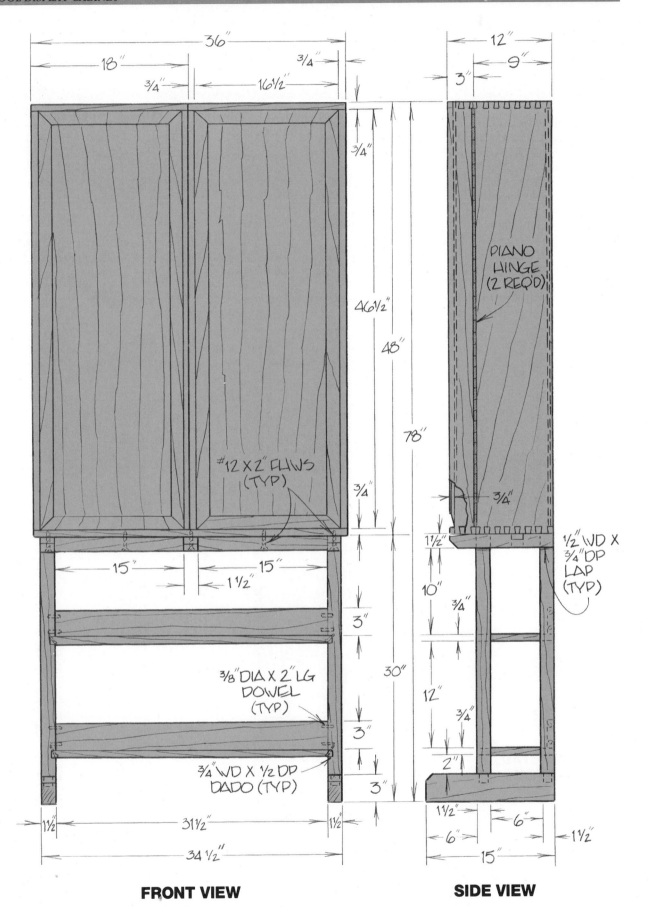

FRONT VIEW   **SIDE VIEW**

4

Cut the case joinery. To make the case as strong as possible, join the top, bottom, and sides with through dovetails. On the cabinet shown, the dovetails were cut with a router and a dovetail jig. (See Figure 1.) If you don't have a jig, you can make these joints by hand with a small saw and a sharp chisel. See Step-by-Step: Cutting Dovetails by Hand for instructions.

After cutting the dovetails, rout ¼"-wide, ⅜"-deep double-blind grooves in the inside surface of the top, bottom, and sides, near the back edges. These grooves hold the plywood back. Each groove stops about ¼" from either end of the board. You don't want to see the grooves in the assembled cabinet.

Also rout the blind dadoes in the sides, as shown in the *Side Layouts,* and a blind dado in the bottom. These

1/There are two router jigs available commercially that will cut through dovetails: the Leigh Dovetail Jig and Keller Dovetail Templates. (Shown is the Leigh Jig.) Both are available through the mail from many woodworking suppliers.

joints hold the shelves, partitions, and lower drawer divider. Each dado is open to the front edge, but it stops at the back groove.

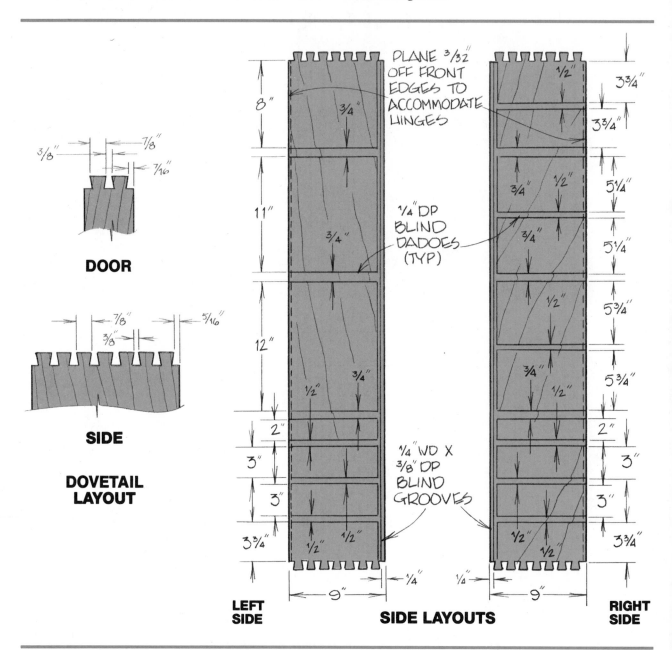

DOOR

SIDE

DOVETAIL LAYOUT

PLANE ³/₃₂" OFF FRONT EDGES TO ACCOMMODATE HINGES

¼" DP BLIND DADOES (TYP)

¼" WD X ⅜" DP BLIND GROOVES

LEFT SIDE

RIGHT SIDE

SIDE LAYOUTS

5

Assemble the case. Finish sand all case parts, then dry assemble them to check the fit of the joints. Make sure the front edges of the top, bottom, and sides are flush. When you're satisfied the parts go together properly, take them apart and rip or plane ³⁄₃₂" from the *front* edge of both *sides*. This will create recesses for the door hinges in the completed case.

Reassemble the case, gluing the dovetail joints *only*.

Don't glue the back in place; let it float in the grooves. Check that the case is square as you clamp the parts together, and wipe away any glue that squeezes out of the dovetail joints with a wet rag.

After the glue dries, sand all joints clean and flush. Also sand any surfaces that you wiped with the rag. This will smooth the raised grain and remove any glue residue that may remain.

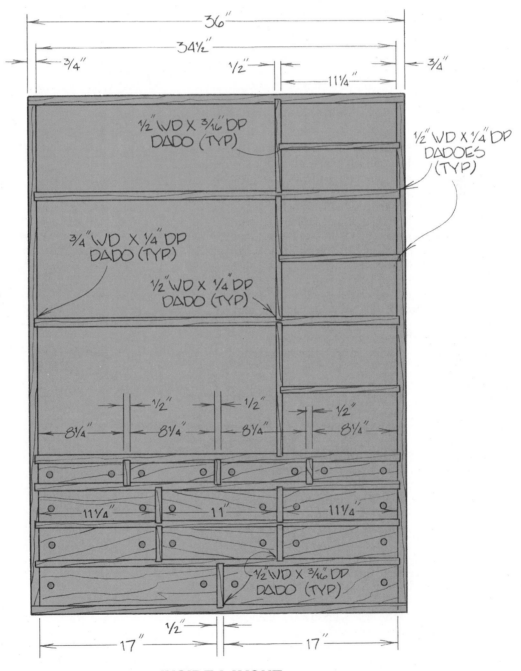

INSIDE LAYOUT

Making the Stand

6 **Cut the stand parts.** After you make the case, build a stand to put it on. The stand will hold the case at a comfortable height, making it easier to fit the shelves, drawers, and doors.

Carefully measure the bottom of the case to see if the dimensions have changed from the plans. If they have, make corresponding changes in the dimensions of the stand and cut the stand parts. Chamfer the upper front corners of the feet and the bottom front corners of the supports. Taper the upper edge of the supports as shown in the *Support Detail*.

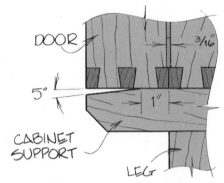

SUPPORT DETAIL

7 **Cut the stand joinery.** Most of the stand parts are assembled with lap joints and dadoes. Only two parts are joined differently: The legs are attached to the feet with mortises and tenons, and the backstops are attached to the legs with dowels.

Cut the dadoes and laps first, using a table-mounted router or a dado cutter. You can also make the tenons on the bottom ends of the legs with the same tools. To make the mortises, drill 1"-diameter, 1"-deep holes in the feet, then square the corners with a chisel.

Dry assemble the stand and carefully mark the locations of the dowels on both the backstops and the legs. Disassemble the stand and drill ⅜"-diameter, 1"-deep holes for the dowels. If you have a plate joiner, you can attach the backstops with wooden plates instead of dowels.

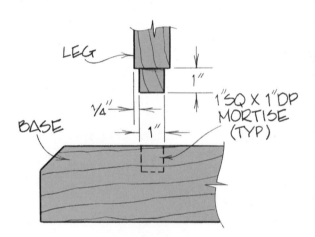

LEG-TO-BASE JOINERY

8 **Assemble and attach the stand.** Finish sand the parts of the stand, being careful not to round any adjoining corners. Also be careful not to remove too much stock with the sandpaper. If you do, the lap and dado joints won't fit properly.

Assemble the stand with glue, wiping off any excess with a wet rag. Reinforce the dado joints that hold the bottom shelf with #10 x 1¼" flathead wood screws. Drive the screws at an angle up through the shelf and into the legs, as shown in the *Shelf-to-Leg Joinery Detail*.

After the glue dries, sand all joints clean and flush. Then attach the stand to the case with #12 x 2" flathead wood screws, driving the screws up through the supports and the stretcher, and into the bottom of the case.

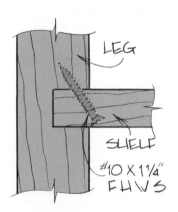

SHELF-TO-LEG JOINERY DETAIL

Making the Shelves

9 **Cut and fit the shelves and partitions.** After you mount the case on the stand, measure the inside to see if the dimensions have changed from the plans. If they have, make corresponding changes to the dimensions of the long shelves and partitions — the *long, horizontal* parts inside the case. Cut these parts to size. Fit them into the dadoes inside the case, temporarily sliding them in place.

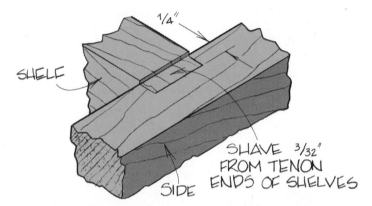

1/4"

SHELF

SHAVE 3/32"
FROM TENON
ENDS OF SHELVES

SIDE

SHELF-TO-SIDE JOINERY DETAIL

10 **Cut the dadoes in the shelves and partitions.** With the long shelves and partitions in the case, measure and mark the dadoes for the vertical dividers. Remove the horizontal parts from the case and cut the dadoes. Finish sand the parts, then glue them in their respective dadoes. Once again, wipe away excess glue with a wet rag.

11 **Cut and fit the dividers and short shelves.** Cut and fit the dividers (*vertical* parts) in the same manner as the long shelves and partitions. When the dividers are all in place, measure and mark the dadoes for the short shelves. Cut these dadoes, then cut and fit the short shelves. Finish sand the parts you have just made and glue them in place.

Making the Drawers

12 **Cut the drawer parts.** Measure the drawer openings and note any deviation from the plans. Cut the drawer parts so each drawer will be about 1/16" *wider* and *taller* than the drawer opening. This will allow you to sand each assembled drawer to size, custom-fitting it to its opening.

13

Cut the drawer joinery. The sides of each drawer are joined to the front and back with lock joints (sometimes called dado-and-tongue joints). You can make these joints on a table saw with an ordinary combination blade that cuts a ⅛"-wide kerf, as shown in Figures 2, 3, and 4.

After making the lock joints, cut ¼"-wide, ¼"-deep grooves in the inside surfaces of all the parts to hold the drawer bottoms. Each groove must be ¼" from the bottom edge, as shown in the *Drawers/Side View* drawing.

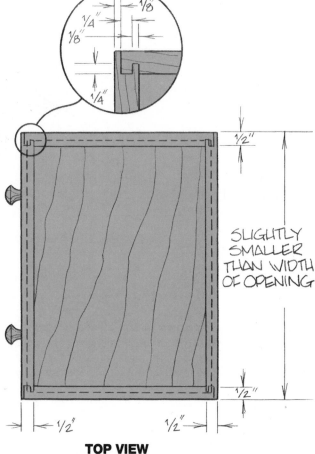

2/To make a lock joint, first cut a dado on the inside face of the drawer side. Use a table saw and an ordinary combination blade — the kerf will be exactly ⅛" wide.

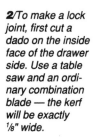

3/Cut a groove in the end of the drawer front, still using the combination blade, in two passes. Feed the ½"-thick workpiece over the blade, making a ⅛"-wide kerf, ⅛" from the face of the board. Turn the board edge-for-edge and make another pass, widening the kerf to ¼".

TOP VIEW

*4/The ¼"-wide kerf, centered in the end of the drawer front, will create two ⅛"-wide tenons — one on either side of the kerf. Trim the **inside** tenon so it fits the dado in the drawer side.*

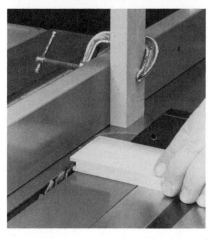

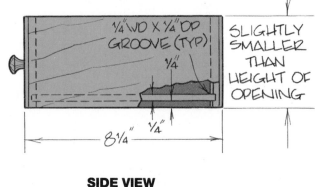

SIDE VIEW

DRAWERS

14 Assemble and fit the drawers.

Assemble the drawers with glue, wiping away the excess. After the glue dries, sand all joints clean and flush. Carefully fit each drawer to its opening, planing, scraping, and sanding away stock until the drawer slides in and out of the case smoothly. If you accidentally slide a drawer too far into an opening to pull it out, use a piece of masking tape as a temporary drawer pull.

Note: Wood expands and contracts with changes in temperature and humidity. You should fit the drawers snugly if you make them in the summer, when it's hot and humid, because the wood will have expanded. Allow just a 1/64" to 1/32" gap on all sides. When the wood shrinks in the winter, the drawers will still operate smoothly. If you make the drawers in the winter, fit the drawers with a little play — about a 1/16" gap all around — to allow the wood room to expand when summer comes.

With the drawers in place, mark the positions of the pulls on the drawer faces. Drill pilot holes in the faces and install the pulls.

Making the Doors

15 Cut the door parts.

Carefully measure the front of the case, and make any necessary changes in the dimensions of the door parts. Like the drawers, the doors should be made slightly larger than the case and then planed to fit. However, don't make them quite as oversized as you made the drawers. Cut the parts just 1/32" longer than needed.

16 Cut the door joinery.

Join the parts of the door in the same way that you joined the case. Cut through dovetails to join the door sides, tops, and bottoms, and blind grooves to hold the door panels. Rip or plane 3/32" off the *back* edge of the *outside* door sides. As mentioned in Step 5, this will make recesses for the door hinges.

17 Cut the raised door panels.

The door panels are raised, as shown in the *Panel Profile*. There are several ways to raise these panels. The most common method is to bevel-cut the panel edges and ends with a table saw. Use the rip fence and a tall fence extension to guide the stock, and adjust the depth of cut so the saw leaves a 1/8" step between the flat and the beveled areas of the panel. (See Figure 5.)

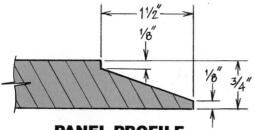

1½"
1/8"
1/8" 3/4"

**PANEL PROFILE
SAW-CUT VERSION**

5/When cutting a raised panel with a table saw, adjust the angle of cut to approximately 15°. Saw the bevels with a hollow-ground planer blade so the cuts will be smooth and clean.

You can also cut raised panels with a panel-raising router bit or shaper cutter. The panels shown were made with a router bit mounted in a drill press. (See Figure 6.)

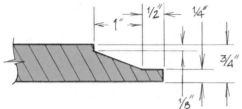

PANEL PROFILE
ROUTER-CUT VERSION

6/If you can adjust your drill press to run at high speeds (5,000 rpm or above), use it as an overhead router to make the raised panels. Mount a raised-panel bit in the chuck and cut the panels in multiple passes. Lower the bit ⅛" to ¼" with each pass, until the lip of the panel is ¼" thick.

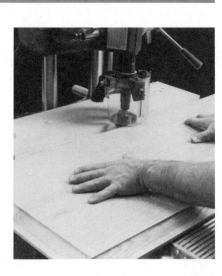

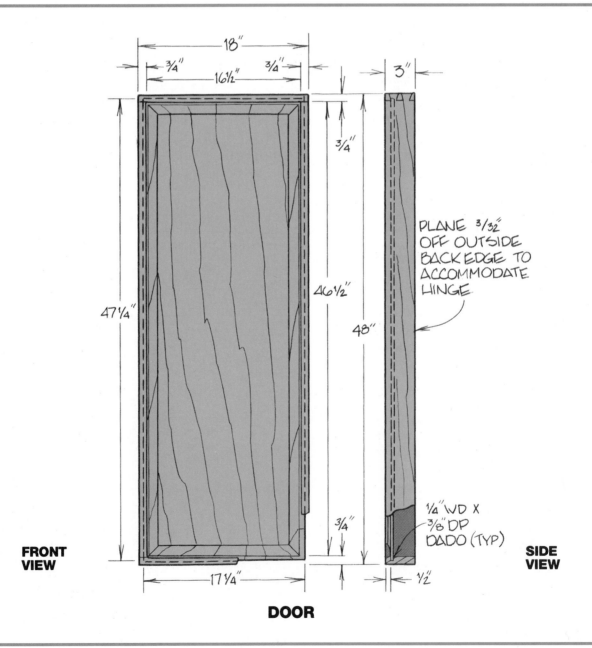

PLANE ³/₃₂"
OFF OUTSIDE
BACK EDGE TO
ACCOMMODATE
HINGE

¼" WD X
³/₈" DP
DADO (TYP)

**FRONT
VIEW**

**SIDE
VIEW**

DOOR

18 **Assemble and mount the doors.** Dry assemble the door parts. Check the fit of the joinery *and* the fit of the doors to the cabinet. When you're satisfied everything fits properly, take the doors apart and finish sand the parts. Reassemble the doors permanently, gluing the dovetails. Don't glue the panels in place; let them float in the grooves so they can expand and contract freely.

When the glue dries, sand all joints clean and flush. Mount the doors to the cabinet with piano hinges. If necessary, scrape or sand the outside surfaces of the doors to make them flush with the cabinet.

Finishing Up

19 **Make the hangers for the doors.** The hangers in the doors must be custom-designed for your tools. The drawings show three possible types of hangers — a *Tool Rack,* a *Saw Holder,* and a *Holder for Hammers, etc.* Mount long, slender tools — screwdrivers, chisels, awls, and so on — in a rack. This is just a horizontal board with holes drilled through it. To make it easier to fetch and replace these tools, cut a narrow slot from the front edge of the board to the circumference of the holes. Hang hammers, saws, and similar tools on pegs or blocks, and keep them in place with catches or turn buttons. Space the pegs and shape the blocks to fit the individual tools.

If you want to change the configuration of racks and holders from time to time, attach them to the insides of the doors with screws, *not* glue. When you replace or add tools, unscrew the racks and holders, rearrange them, and tighten them down again.

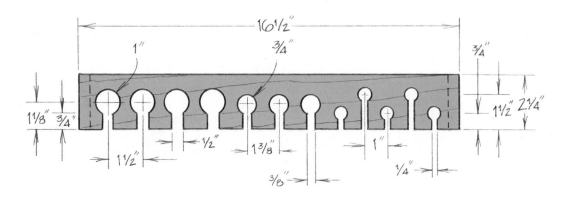

TOP VIEW

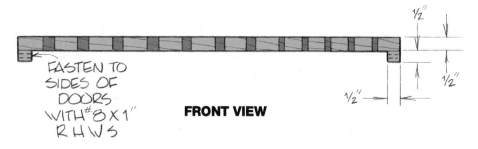

FRONT VIEW

TOOL RACK

20

Finish the cabinet. Remove the doors and drawers from the cabinet, and detach the case from the stand. Remove all the hardware and set it aside. Do any necessary touch-up sanding, then coat all wooden surfaces with a *penetrating* finish — something that soaks into the wood and protects it from the inside out. (Finishes that build up on the surface are too easily scratched by the metal tools.) The cabinet shown is finished with a mixture of about 75 percent tung oil and 25 percent spar varnish, applied with a cloth. This makes a durable, waterproof finish.

Apply several coats, taking care to cover the inside of the cabinet as many times as the outside. If the parts are finished unevenly, they may warp or cup. After the finish dries, buff it with #0000 steel wool and paste wax until the surface takes on a smooth, deep gloss. Finally, reassemble the cabinet and the hardware.

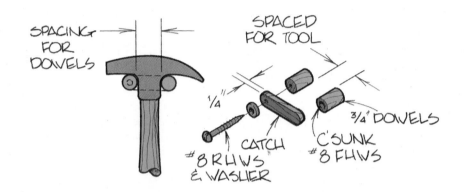

HOLDER FOR HAMMERS, ETC.

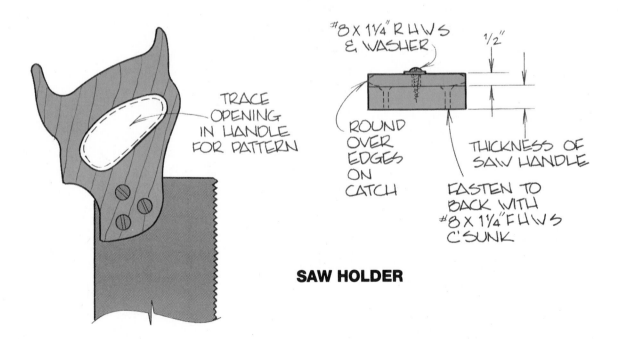

SAW HOLDER

Step-by-Step: Cutting Dovetails by Hand

Handmade dovetails have always been a mark of fine furniture. When properly made, these joints are stronger and more durable than any other joint in woodworking. They also add grace and elegance to a project. It's a small wonder that contemporary furniture designers sometimes display dovetail joinery on the *outside* of their pieces.

Despite their strength and handsome appearance, dovetails are not difficult or expensive to make. They don't require special tools — just a small saw, a sharp chisel, and a simple jig you can make from scraps. Nor do they require any special skills. Just patiently and carefully follow this procedure:

Always begin a dovetail joint by making the *pins.* Lay them out on the end of a board. The spacing of the pins is not critical, but the angle of the joint should be between 7° and 10° (off perpendicular to the face of the board). Cut the lines with an awl or marking knife; these lines are harder to see than pencil lines, but they're more precise. They also help to prevent the wood from chipping or splintering when you cut the joint.

Saw the sides of the pins with a small saw. Both a dovetail saw and a Japanese *dozuki saw* work well. To make sure that you stop sawing at the base of the pins, make a U-shaped hardwood jig as shown. Clamp this jig to the board, even with the base of the pins. Stop the saw just as the teeth touch the arms of the U.

Remove the waste between the pins with a very sharp chisel. First, use it as a cutting tool. Hold the chisel *vertically* and place the edge on the line at the base of the pins. Strike it with a mallet, driving it about 1/8" into the wood.

Then use the chisel as a wedge. Hold it *horizontally* at the tops of the pins. Place the edge about 1/8" below the surface. Lightly tap the chisel, prying out a 1/8"-thick piece of waste. Repeat, using the chisel alternately as a cutting tool and a wedge until you have cut halfway through the waste. Then turn the board over to remove the remaining waste.

Step-by-Step: Cutting Dovetails by Hand — Continued

5

Use the completed pins as a template to lay out the tails on the adjoining board. Once again, use an awl or a marking knife to make the lines.

6

Saw the sides of the tails as you did the sides of the pins. If you have trouble keeping the saw at the proper angle, use the U-shaped jig as a guide. Clamp it to the board, even with the side of the tail you want to cut. Let the flat side of the saw rest against the jig as you work.

7

Remove the waste between the tails with a chisel, again using the chisel alternately as a cutting tool and a wedge.

8

Test fit the joint. If it's too tight, shave a little stock from the pins. If it's too loose, glue tiny pieces of veneer to the pins. The grain direction and the species of the veneer should match that of the pin board. Properly done, this fix will be invisible.

Credits

Contributing Craftsmen and Craftswomen:

Richard Belcher (One-Board Cases)

Mark Burhans (Glass-Top Display Stand)

Dan Callahan (Multiple Picture Frame)

Nick Engler (Stackable/Hangable Shelves, Quick-and-Easy Picture Frames, Tool Display Cabinet, Corner Showcase, Watch Case)

Mary Jane Favorite (Round and Oval Frames, Corner Cupboard, Country Shadowboxes)

W.R. Goehring (Glass-Front Wall Cabinet)

Bob Hines (Quilt Rack)

Fred Matlack (Pigeonholes)

John Shoup (Curio Cabinet)

Note: One project, the Museum Table, was built by a craftsman whose name has been erased by time. We regret that we cannot tell you who built it; we can only admire his (or her) craftsmanship.

The designs for the projects in this book (with the exception of the Museum Table) are the copyrighted property of the craftsmen and craftswomen who built them. Readers are encouraged to reproduce these projects for their personal use or for gifts. However, reproduction for sale or profit is forbidden by law.

Special Thanks To:
Larry Callahan
Gordon Honeyman
Chip and Mary Anne Marderosian
Patterson Flowers, West Milton, Ohio
Shopsmith, Inc., Dayton, Ohio
Wertz Hardware Store, West Milton, Ohio

Rodale Press, Inc., publishes AMERICAN WOODWORKER™, the magazine for the serious woodworking hobbyist. For information on how to order your subscription, write to AMERICAN WOODWORKER™, Emmaus, PA 18098.

WOODWORKING GLOSSARY

Parts of a Board

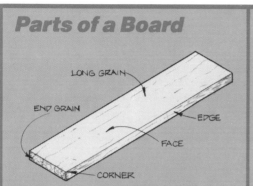

LONG GRAIN
END GRAIN
EDGE
FACE
CORNER

Basic Saw Cuts

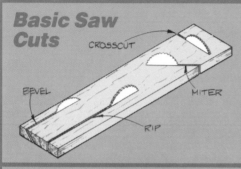

CROSSCUT
BEVEL
MITER
RIP

Parts of a Drawer

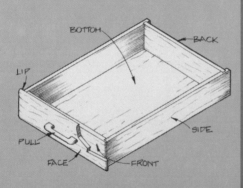

BOTTOM
BACK
LIP
SIDE
PULL
FACE
FRONT

Parts of a Frame

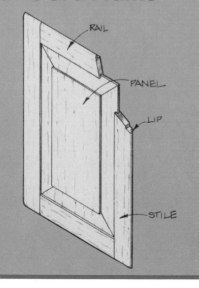

RAIL
PANEL
LIP
STILE

Basic Joinery

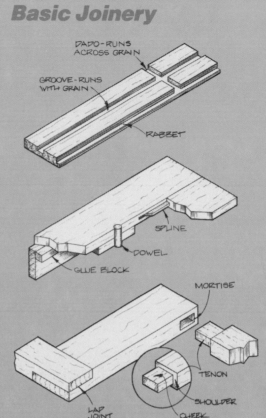

DADO-RUNS ACROSS GRAIN
GROOVE-RUNS WITH GRAIN
RABBET
SPLINE
DOWEL
GLUE BLOCK
MORTISE
TENON
SHOULDER
LAP JOINT
CHEEK

Common Shapes and Moldings

QUARTER-ROUND
BEAD
OGEE, OR CYMA CURVE
CORNER
COVE
BED
CROWN

CABRIOLE
TAPER
STRAIGHT

Holes

SCREW HOLE
STOPPED HOLE
THRU HOLE
COUNTERBORE
COUNTERSINK
PILOT HOLE

Parts of a Ta[ble]

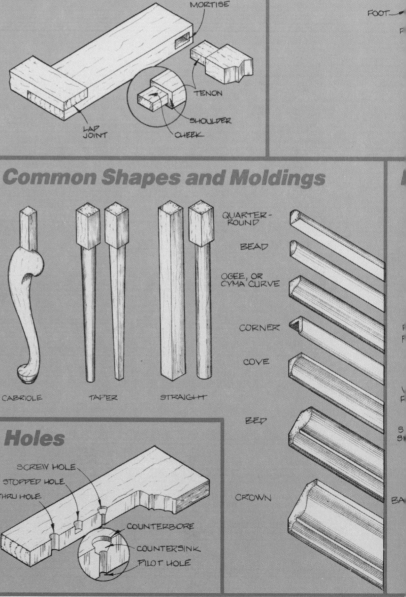

LEAF
LEG
KNEE
ANKLE
FOOT